TOP 100 HIGHWAY CODE QUESTIONS AND ANSWERS

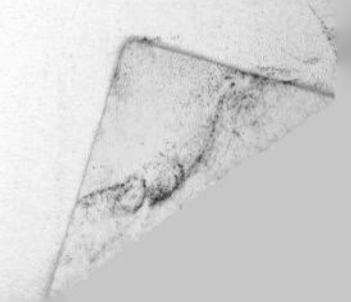

TOP 100 HIGHWAY CODE QUESTIONS AND ANSWERS

Gary Endenwall

foulsham
LONDON • NEW YORK • TORONTO • SYDNEY

foulsham
Bennetts Close, Cippenham, Berkshire SL1 5AP

ISBN 0-572-01917-3

Phototypeset in Great Britain by Typesetting Solutions, Slough, Berks.
Printed in Great Britain by Cox & Wyman Ltd., Reading, Berks.

Contents

Introduction

At the end of your driving test the examiner will ask you a number of questions about the Highway Code. If you get the answers wrong, you will probably fail the test — no matter how well you have driven. Rule Number One is: *learn your Highway Code.*

But the code is not just something to help you get a driver's licence. It is a set of rules and recommendations aimed at helping everyone — learners, cyclists, pedestrians and experienced drivers — to behave properly and safety on the road. In fact the Road Traffic Act of 1972 spells out the importance of knowing the code well:

"A failure on the part of a person to observe a provision of the Highway Code shall not of itself render that person liable to criminal proceedings . . ., but any such failure may in any proceedings . . . be relied upon . . . as tending to establish or to negative any liability which is in question in those proceedings."

Rule Number Two is: *don't forget your Highway Code* — read it through regularly.

This book will help you to learn the code and to keep it fresh in your mind. Ask a friend to read out the questions and check your answers.

Diagrams have been adapted from *The Highway Code* with the permission of the Controller of Her Majesty's Stationery Office.

Roads are pretty full these days, with many different kinds of users sharing a relatively narrow space. Selfishness or carelessness from any one of them can endanger others.

LANE 1

Pedestrians & Cyclists

Not everyone drives a car. But all of us, at some time or another, cross or walk on a road. Pedestrians are far more vulnerable than car drivers, because they are not protected by a metal box around them. Furthermore, their thoughtless actions can cause serious accidents, even if they themselves escape unhurt. An ill-timed dash into the street can make a driver swerve or skid. So even shoppers must know their Highway Code.

The same goes for cyclists. By and large, cyclists are treated very badly by motorists. They are either ignored altogether, or looked on simply as nuisances. Some cyclists encourage these views by failing to observe strict road discipline. You don't have to pass a test to ride a push-bike on the road, but you certainly should know the rules.

Picture Quiz: Right or Wrong?

QUESTION 1: This is a road on the edge of a country town. Which of the pedestrians shown is walking correctly?

QUESTION 2: What road sign would you expect to see at or before the point where the road bends right (E)?

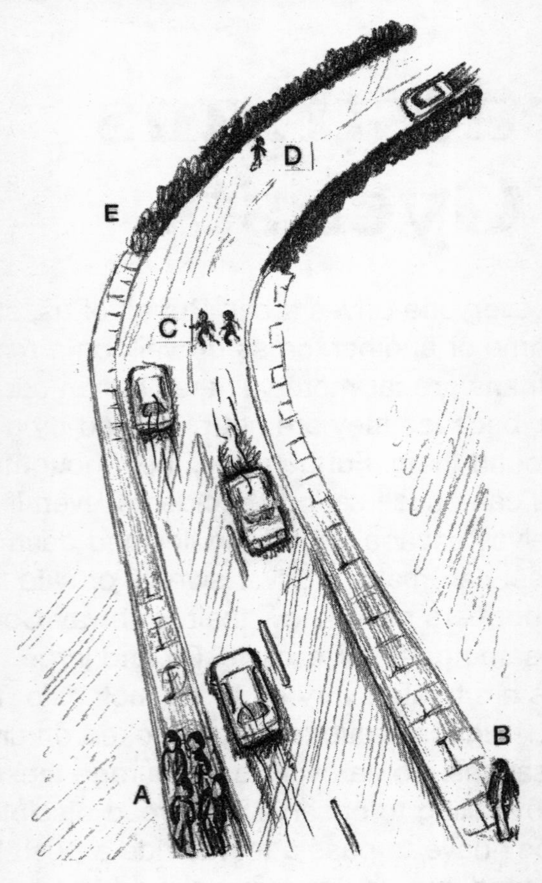

(Answers: Page 188)

(i) Rules for pedestrians

Questions

1 Where is the safest place to walk on a road?

2 What should I do before I step into the road?

3 If there is no pavement, on which side of the road should I walk?

4 How should two or more people organise themselves when walking along a road without a footpath?

Answers *(Rules for pedestrians)*

1 On a pavement or footpath. Where possible avoid walking next to the kerb with your back to the oncoming traffic.

2 Look both ways for traffic.

3 On the right-hand side of the road. This way you will be facing the oncoming traffic. Keep close in to the side of the road, especially on right-hand bends where a car driver may not be able to see you until the last moment. It may be safer to cross the road well before you reach a sharp right-hand bend so that

oncoming traffic has a better chance of seeing you. Then, after the bend, cross back to face the oncoming traffic.

4 They should go in single file and never side-by-side. Always leave as much room as possible for vehicles to get by. Buses and articulated lorries take up a lot of space.

If you are part of a very large group, however, it is best to keep to the left-hand side of the road in a compact mass. Look-outs should be appointed — one to keep watch at the front and one at the back.

5 Should a pedestrian wear any special kind of clothing?

6 Is fluorescent clothing useful at night as well?

7 Can I walk on motorways or their slip roads?

8 How should I walk with young children on the road?

9 On a tram route, how should I cross the road?

Answers

5 Yes. It is important that you wear clothes which are bright or at least light in colour. Dark clothing tends to merge into the background

and makes you hard to see. The most conspicuous clothes are made of fluorescent material which seems to glow in daylight. If you are not wearing any bright or light clothes, then at least carry a bag or coat which can easily be spotted.

6 No. It does not show up well in the dark. Instead try and wear something shiny which will reflect the light from car headlights. Best of all, carry a torch but don't shine it directly in the driver's eyes. Large groups should carry a white torch at the front and a red one at the rear. People walking on the outside should also carry lights and wear reflective clothing.

7 Certainly not, except in the case of a vehicle breakdown (see page 100). Pedestrians are not allowed on motorways, even for the purpose of thumbing lifts. Nor are drivers allowed to pick up or set down passengers on motorways or slip roads.

8 Always hold their hands and walk between them and the traffic. Never let them run out into the road or go on the road alone. Very young children should be strapped in push-chairs or kept on reins.

9 In pedestrian areas, the path of the tram will be marked out by shallow kerbs, changes in the road surface or white lines. If the track is unfenced you may cross at any point. Look both ways and when it is clear walk straight across; do not walk on the track.

Where they are provided, use designated crossing places. Some may have flashing amber lights to warn you that a tram is approaching. Do not start to cross the track when the lights are flashing. If you are already crossing when the lights start to flash, it will be safe to continue. Avoid treading on the rails.

Quiz: Who Carries What?

A. Pedestrians; **B.** Cyclists; **C.** Motorcyclists; **D.** Road menders; **E.** Traffic crossing wardens.

(Answers: Page 188)

(ii) Crossing the road

Questions

1 What is the Green Cross Code?

2 What are the rules of the Code?

3 Is it safe to cross a road with an island in the middle?

4 What do the zigzag lines near a Zebra crossing mean?

5 Does a pedestrian have a right of way over traffic at a Zebra crossing?

Answers *(Crossing the road)*

1 It is a set of simple rules for pedestrians who want to cross a road. Children should be taught the Green Cross Code as soon as they are able to understand it.

2 There are six basic rules.

(a) Find a safe place to cross. The safest places of all are subways. Pelican or Zebra crossings, traffic lights or crossing patrolled by 'lollipop' men or women.
Always choose a clear spot where drivers can easily see you. Never cross between parked cars.

(b) Wait on the kerb — but not too near the edge of the road. Make sure that you can see clearly any approaching traffic.

(c) Look and listen. Look both to the right and to the left. And keep your ears open — an unseen vehicle may be approaching at high speed.

(d) Wait until the traffic is past. Then look both ways again to check that you haven't missed anything.

(e) When there is no traffic near, walk across the road. Walk — don't run — and go straight across without lingering.

(f) Keep on looking and listening. Keep your wits about you once you are on the road. Traffic may still appear suddenly.

3 Yes, and in many ways it is safer. You will only have to worry about one set of vehicles at a time. Use the Green Cross Code to reach the island. Then stop, look again (both ways) and use the code to cross the other lane.

4 They are there to warn drivers and pedestrians that a Zebra crossing is near. They also mark the danger zone near a crossing. You must always use a Zebra crossing if it is nearby and never cross a short distance away.

5 Only when he or she has stepped onto the crossing. Until a foot is placed on the crossing

the traffic is not obliged to stop. Even then pedestrians should give vehicles plenty of time to notice them and to slow down and stop. Never push a wheelchair or pram onto a crossing until the traffic has stopped.

6 What is the sequence of lights for pedestrians at a Pelican crossing?

7 Am I allowed to begin crossing at a Pelican crossing, are the lights for each lane operated separately?

8 If there is an island in the middle of a Pelican crossing, are the lights for each lane operated separately?

9 What is the difference between a Pelican and a Puffin crossing?

10 How do I cross at a Puffin crossing?

Answers

6 The first light shows a red man. This means that you must wait, and press the button on the box. When the lights change, a green man is shown. This means that you may cross. There may also be a bleeping noise to alert blind pedestrians. After a few seconds the green man signal begins to flash, warning you to hurry. Then the red man is shown again.

7 No. The flashing shows that the lights are about to change. You should not start to cross.

8 Only if the crossings are staggered (not in a straight line). When you reach the island you must press the button and wait for the green man signal to appear. Then walk across.

9 Puffins are very similar to Pelican crossings except that the pedestrian signals are on your side of the road.

10 After pressing the button, wait where indicated for the green figure to show. Infra-red detectors will vary the length of time that the red light stays on for drivers to ensure that pedestrians have enough time to cross safely.

11 How should I cross at traffic lights if there are no pedestrian signals?

12 If there is a police officer or a school-crossing patrol directing the traffic, when should I cross the road?

13 What is the purpose of guard rails between pavement and road?

14 Why do some pedestrian crossing points have textured paving?

15 What should I look out for when crossing a one-way street?

Answers

11 Make sure that you are standing at a spot where you can see the lights. Keep a close watch on the traffic. Never try to cross when the lights are green. Even when the lights turn red, check that there are no filter lanes.

12 When the officer signals that you should cross. Walk in front of him — never behind.

13 They are there to prevent pedestrians from walking on dangerous parts of the road, where vehicles may be travelling at high speeds. You should not climb over them, but cross only where there are gaps in the rails.

14 Textured paving shows blind or partially-sighted people where to stand while waiting to cross the road.

15 Beware of one-way streets! There may be more than one lane, so look carefully at the road markings to see how many lanes there are. Vehicles in one-way streets may also be travelling faster than they would in normal circumstances.

Check which way the traffic is moving. In some one-way streets bus lanes operate in the opposite direction to the rest of the traffic. Do not cross until it is safe to do so.

16 Why should I take special care when crossing a bus lane?

17 I know that it is dangerous to cross a road from between two parked cars. But what if I have no choice?

18 What precautions should I take when crossing a road at night?

19 What should I do if I hear an ambulance coming?

20 How should I board a bus at a request stop?

Answers

16 Bus lanes are there for one reason, so that buses and only buses can travel freely along them. Because there are no obstructions, the buses may be travelling fast. And remember that bus lanes sometimes run against the flow of the traffic. Check very carefully which way the buses are running before you step onto a bus lane. The same rules apply for cycle lanes.

17 Walk between the two cars and stop at the outside edge. Then look to make sure that you can see clearly in both directions and that the drivers can see you. Also make sure that neither of the parked cars has its engine running. It may be just about to move off, and the driver may not have spotted you.

18 Make sure that drivers are able to see you. First of all look to see if there is a Pelican or Zebra crossing and use that. If not, cross near a street light. When going out at night, wear clothing which is bright or light in colour. Best of all, wear something which is shiny and reflects light.

19 If you see or hear any emergency vehicle approaching (an ambulance, a fire engine or a police car), keep well out of the way. The emergency vehicle will be in a hurry and other cars will be trying to let it through by moving to the side of the road. Stay firmly on the pavement.

20 First of all make a clear signal to the driver to show that you want him or her to stop. Then, when the bus has pulled in, wait until it has stopped before you start to climb on board. As a general rule, you must never get on or off a bus unless it is at an official bus stop.

(iii) Rules for cyclists

Questions

1 Which parts of a bicycle should I check before I set off for a ride?

2 What procedure should I follow when moving away from the kerb?

3 Should cyclists ride side-by-side or in single file?

4 Are cyclists allowed to ride on pavements or footpaths?

5 I want to turn right, but the road is very busy. What should I do?

Answers *(Rules for cyclists)*

1 The tyres should be firmly pumped up. Look all round them to be sure that they are not badly worn or scuffed. The chain should have a small amount of slack in it, and should be lubricated regularly with grease or oil. The brakes should work efficiently and the saddle should be the right height. Check the brake blocks for wear. If you are riding at night, you must have a rear reflector and front and rear lights which work properly. Fit a bell so that you have some way of warning other road

users, particularly blind and partially-sighted people, that you are there.

2 Look behind you. You should do this before you make any sort of manoeuvre on the road. If there is no vehicle coming, put out your arm to indicate that you are pulling out. Then ride steadily on. But be careful that there are no bumps or potholes in the road which will force you to swerve before you have picked up speed. Take care near road humps, narrowings and other traffic calming features.

3 Two cyclists can ride side-by-side as long as the road is wide enough and there is plenty of room for vehicles to overtake them in comfort. But never ride three abreast. On narrow roads and cycle lanes, ride in single file.

4 Only when there are signs specifically allowing you to do so. Remember that pedestrians have the right of way. Use cycle lanes and tracks where possible.

5 Play safe and pull into the left-hand side of the road first. Stop there and wait until there is an opportunity for you to cross over to the right. Make sure that there is no fast-approaching traffic. It may be safer to wait on the left until there is a safe gap, or to dismount and walk your cycle across the road.

6 How should a cyclist proceed at a roundabout?

7 What are the main rules for safe cycle-riding?

8 What is the difference between a cycle lane and a cycle track?

9 What special care should I take at (a) road junctions and (b) dual carriageways?

Answers

6 If you feel confident enough, you should proceed in the same way as any other road vehicle (see page 85). However, big roundabouts can be confusing and frightening places for a cyclist. The simplest course is to stick to the left-hand lane until your reach your turn-off. Beware of other vehicles cutting across you to turn off. If you have any doubts, simply get off the bike, find a pavement and walk.

7 Remember the following points:

- Keep both hands on the handlebars and both feet on the pedals.
- Never ride close behind another vehicle and certainly do not attempt to 'hitch a lift' by holding onto another vehicle or cyclist.
- Never carry another person or any awkward baggage which might get tangled in the wheels or cause you to overbalance.

- Never lead an animal while you are cycling.
- Wear a cycle helmet which conforms to recognised safety standards.
- Always wear light-coloured or reflective clothing. Avoid long or loose clothing that may get tangled in the wheels.
- Always use a cycle path if there is one available.
- Remember that traffic signals also apply to cyclists. You must not cross the stop line across the road when the lights are at red. Some junctions have advanced stop lines which enable cyclists to position themselves ahead of other traffic. Where these are provided, use them.
- You must not ride under the influence of drink or drugs.
- Do not leave your cycle on any road where it is likely to cause danger to other road users, or where waiting is prohibited.

8 Cycle lanes are marked by either an unbroken or broken white line along the carriageway. Keep within the lane and watch out for traffic emerging from side turnings.

Cycle tracks are located away from the road. Where a cycle track is shared with a footpath, you *must* keep to the track intended for cyclists. Watch out for pedestrians, especially the elderly and people with disabilities.

Cycle tracks on opposite sides of the road are sometimes linked by signalled crossings. If the crossing is provided for cyclists only, you may ride across but you *must not* cross until the green circle symbol is showing. Do not ride across a Pelican crossing.

9 (a) At road junctions cyclists should watch out for vehicles turning in front of them from or into a side road. Do not overtake on the left of vehicles slowing down to turn left. Pay particular attention to long vehicles, particularly at corners where they may have to move over to the right before turning left. Wait until they have turned before moving on.

(b) If you are crossing or turning on to a dual carriageway where there are no traffic light signals, great care must be taken. Wait for safe gaps and cross each carriageway in turn; remember that the traffic will be travelling quickly.

QUESTION 1: What sound does this emergency vehicle make as it speeds along a road?

QUESTION 2: If you see or hear an emergency ambulance passing in your direction, what might this tell you about the road ahead?

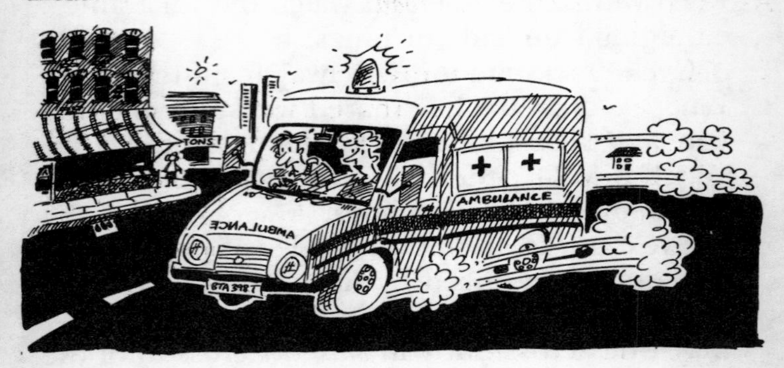

(Answer Page 188)

(iv) Animals on the road

Questions

1 What sort of problems can be posed by a dog in a car?

2 What is the most important piece of clothing a horse rider must wear on the road?

3 Which side of the road must a horse rider keep to?

4 What sort of clothing should I wear if I am riding or herding animals after sunset?

Answers *(Animals on the road)*

1 There are two main problems. One is that the dog may be badly controlled in the car and distract the attention of the driver. In this case the dog should be kept strictly under control. The second problem is that the dog may dash into the road as soon as a door is opened. The answer to this is to put it on a lead before you open the door. Never let a dog out onto the road without a lead.

2 A hard hat; children under the age of 14 *must* do this. Head injuries are common and can be serious.

3 To the left. If you have dismounted and are leading the horse you must also keep to the left-hand side of the road, and keep the animal on your left. Horses must not be ridden or led along footpaths or pavements at the side of the road. Never ride more than two abreast; ride in single file on narrow roads.

4 The same rule applies as for walkers and cyclists; wear light-coloured or reflective clothing; your horse should also have reflective bands above the fetlock joints. In the dark you should also carry a white light at the front and a red light at the back to help other traffic to spot you.

5 I am driving a large herd of cattle along a main road. How should I give warning to other traffic of this slow-moving hazard?

6 What should a horse rider do when approaching an automatic level crossing?

7 What are the main points to remember for safe horse-riding?

8 As a horse rider, can I ride on roundabouts?

Answers

5 Never set out to herd a large number of beasts without getting at least one companion to help you. Send him or her on ahead to stand at awkward spots (bends, the brows of hills) and signal to the traffic to slow down. In some cases they may have to stop. Always keep your herd to the left-hand side of the road if possible. If the herd is too big to manage easily, you will have to split it into smaller groups and drive these along separately.

6 Check well in advance on other traffic and be prepared for the warning to sound. If this happens, stop the horse well back from the crossing. Stay mounted — dismounting will only slow you down later. If all is clear, keep going over the railway.

Road hogs.

7 The main points to remember are:

• Never ride a horse without a saddle or bridle and always make sure all tack fits well and is in good condition.

• Always keep both hands on the reins unless you are signalling; and keep both feet in the stirrups.

• Do not carry another person or anything which might affect your balance or get tangled up with the reins.

8 You should avoid roundabouts if possible, but if you have to use them, keep to the left and watch out for vehicles crossing your path to leave or join the roundabout. Signal right when riding across exits to show you are not leaving. Signal left just before you leave the roundabout.

LANE 2

The Vehicle & its Driver

Most people know very little about how their car works. They can fill up the fuel tank, check the oil level and put air into the tyres, but that's about all. Although it is obviously an advantage to be a skilled mechanic, it is not vital. What is important is that the vehicle is kept in good condition. It should be serviced regularly by a qualified person and given a visual check-over before every journey.

Perhaps someone should check over the driver as well! He or she must also be in good condition if they are to drive safely and responsibly.

Five in One Crossword

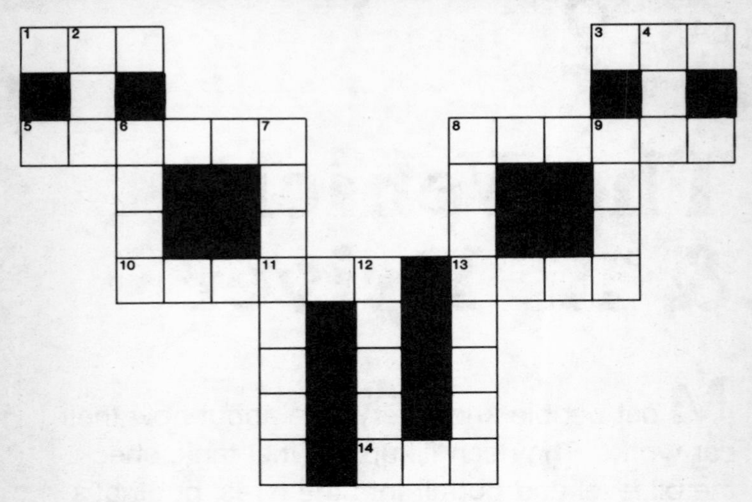

CLUES: ACROSS

1 One car being pulled by another is in ---.
3 Possible punishment for drivers found guilty of drunk or other dangerous driving.
5 Snow, ice or rain leave roads this way.
6 The amount of tread this must have is laid down by law.
8 The oil circulating system of your car's engine.
9 Only a well behaved one should travel as a passenger in a car.
10 In many cities, it was long ago replaced by the 'bus.
11 It helps you find your way when out driving.
13 The MOT is one of them.
14 Number of lights in two sets of traffic lights.

CLUES: DOWN

2 The number of riders a horse should have on the road.
4 Seventeen is the minimum for drivers.
6 Colloquial expression for sounding a car horn.
7 He or she accompanies you on your driving test.
8 Four less than the maximum speed on roads presently allowed in Britain.
9 It can have a lamp on top.
12 They dispense petrol

(Answers: Page 188)

(i) The car

Questions

1 Which parts of the car should I check to help me see clearly on the road?

2 Which parts should I check to help me be seen?

3 Which other parts need special attention?

4 Does the spare tyre have to be as good as the others?

Answers *(The car)*

1 The windscreen, for a start. It should be kept clean of dust and dirt all over. The windscreen wipers and washers should work properly (make sure there is water in the washer fluid reservoir). So must the demister, which keeps the screen clean on the inside. You should also make certain that the other windows, the mirrors, the lights, the indicators, the reflectors and the number plates are clean.

2 All the lights and indicators must work properly. Broken bulbs should be replaced immediately.

3 The brakes and the steering must be in good condition. They are vital for your safety and the safety of others. So are the tyres, which must have the correct amount of tread on them (for the legal requirements, see page 184) and should be pumped up to the pressure recommended by the manufacturer.

4 Yes. A worn tyre should never be kept as a spare. And remember to keep it properly pumped up.

5 Must I wear a seat belt when I am driving a vehicle?

6 Is anyone exempt from wearing seat belts?

7 Does the wearing of seat belts really save lives?

Answers

5 Yes, you must and so must your front-seat passengers, by law. If rear seat belts are fitted, these must also be worn. Make sure that the seat belts lock and unlock properly and that they are adjusted to fit driver and passengers correctly.

6 A few people are exempt. These include delivery drivers on their local rounds and

people who hold a medical exemption certificate. You may remove your seat belt under special circumstances, for instance when you are reversing or making some manoeuvre which involves you in twisting or turning in your seat.

7 Yes. Statistics show that the risk of death or serious injury in an accident is halved by the wearing of seat belts.

8 Am I allowed to drive with a cracked or holed exhaust system?

9 My car is fitted with L-plates. Should they be kept on when the car is being driven by a qualified driver?

10 What precautions should I take if I am carrying a large load?

11 Will tinted glass on the windscreen help me to see better?

Answers

8 No. Your exhaust system must be working efficiently. If there is a hole in the silencer, the noise of the engine explosions will be extremely loud. Get a worn-out system replaced immediately.

9 When the car is not being used for driving practice, the L-plates should be taken off or covered up. This will show others that the driver is not a learner.

10 Never try and carry too much, even if you are using a trailer as well. All loads must be fixed securely and tied if necessary. No part of a load should jut out in such a way as to be dangerous.

11 No, it won't. A clear screen will give you the best vision. All stick-on or spray-on tinting materials should be removed.

Motoring Firsts • 1

Several firsts were scored in the early days of motoring, around the turn of this century. For instance, speeding is not a new offence. Nor is drunk driving. The first driver to be convicted of speeding was Walter Arnold, a miller who was charged with driving over the 2 mph limit in 1896. The first drunk driver was found guilty of being inebriated while driving an electric cab in London in 1897.

The first woman to pass a driving test did so in 1898, the same year the first full-sized petrol driven bus took to the road and Carl Benz fitted one of his cars with the first mileometer.

(ii) The driver

Questions

1 Should I drive if I am feeling tired?

2 Should I drive a car if I am ill or taking medicine?

3 What are the effects of alcohol on the driver?

4 Has alcohol been shown to cause serious accidents?

5 What is the legal limit?

6 What are the penalties for driving above the legal limit?

Answers *(The driver)*

1 Certainly not. If you are tired you will not be able to drive at your best. Your reactions will be slow and your judgement may be faulty. Even worse, you may actually fall asleep. The result could be a serious accident.

2 Illness may have the same effect on you as fatigue (see previous question). It is unwise to drive if you are unwell. Likewise some medicines can make you feel drowsy. Check

with your doctor or the chemist whether your medicine will have this effect.

3 The drinking of alcohol invariably makes you a worse driver. For a start, it gives you a false feeling of confidence. You think that you are driving better than in fact you are. This may lead you to take stupid risks. Alcohol also affects you physically. It slows down the reflexes and makes you late in reacting to events on the road. it decreases the coordination between the brain, hands and eyes, and makes you less able to judge speed and distance correctly. Never drink and drive.

4 Yes. About one in three of all people who are killed in road accidents have more than the legal limit of alcohol in their blood.

5 As measured by a breathalyser, the limit is 35 microgrammes of alcohol per 100 millilitres of breath. However, some drivers may have less alcohol than this in their bodies and still be adversely affected by it.

6 You will lose your driving licence for a long time. You may also be fined and, in very serious cases, given a prison sentence.

7 If I need glasses to correct my eyesight, must I wear them when I am driving?

8 Is it a good idea to wear sunglasses or other tinted glasses?

9 If you are learning to ride a motorcyle or scooter as a learner, what *must* you do?

10 When riding a motorcycle or scooter as a learner, what must you *not* do?

11 What piece of equipment is a motorcyclist legally obliged to wear?

12 What other special clothing should a motorcyclist wear?

Answers

7 Yes. Your eyesight must be up to the standard required for the driving test. During the test the examiner will check your eyesight by asking you to read a nearby number plate at a distance of 21-23 metres, depending on the size of the lettering. So if you need to wear spectacles or contact lenses, you must always wear them when you drive. Otherwise you will be breaking the law.

8 Only in bright sunlight. Tinted glasses, goggles or helmet visors (for motorcyclists) will impair visibility at night or in murky weather conditions.

9 You *must* take basic training with an approved body before riding on the road, unless exempt?

10 You *must not* carry a pillion passenger, pull a trailer or ride a solo motorcycle with an engine capacity in excess of 125cc.

11 Anyone riding a motorcycle, scooter or moped must wear a safety or crash helmet. And the helmet must be fastened securely with a strap or some other device.

12 Strong boots and gloves will give a certain amount of protection against injury and keep the rider warm. It is a good thing to wear bright or light coloured clothing so that other road-users can spot you easily. If possible, wear fluorescent clothing during the day and reflective clothing at night.

13 Where is the safest place in a car for young children to travel?

14 How should very young children be fastened most safely?

15 What should be done if the seat belt cannot be made tight enough to restrain the child?

16 And if there are no seat belts or harness fitted to the rear seat?

17 Is it safe to carry children in the luggage space behind the rear seats?

Answers

13 Young children should travel in the rear seat of the car. They should also wear seat belts or some other kind of fastening.

14 Babies of up to nine months old should be placed in a carry cot strapped to the rear seat and covered with a stiff cover to prevent them from being flung out. Infants who can sit up should be fastened into a special safety seat, facing either backwards or (for toddlers) forwards. Older children should wear a safety harness or an adult seat belt.

15 Safety belts are made for adults, and are of little use on their own for small children. A special booster cushion should be placed behind the child to provide extra bulk. Do not use an ordinary cushion with rounded corners, as this will easily work loose.

16 In this case the child should travel in the front of the car, with the seat belt fastened and, if necessary, a booster cushion. This is safer than leaving a child unrestrained in the back seat. Remember that it is the legal responsibility of the driver to make sure that children under fourteen are properly restrained when they are travelling in the front seat.

17 This is not usually a safe way for children to travel. However, some manufacturers of estate cars and hatchbacks do fit seats in the luggage area for children to sit on.

18 Is it necessary to fit child safety locks on the doors?

19 Am I allowed to speak into a hand-held microphone or telephone headset while I am driving?

20 Am I allowed to stop on the hard shoulder of a motorway to make or answer a telephone call?

21 When I am a qualified driver, can I supervise a learner?

Answers

18 There is no law which makes the fitting of safety locks compulsory. But they are an added safety measure which will stop little hands from opening doors at dangerous moments. If your car has child safety locks, check that they are secured before setting off.

19 You should only use hand-held equipment when the vehicle is not moving. It isn't safe to drive and make a telephone call at the same time. If you must use a car telephone speak into a microphone which is slung round the neck or clipped onto the clothing, otherwise you may be distracted from what is happening on the road.

20 No.

21 You may supervise a learner driver in a car *only* if you are at least 21 years old and have held a full British licence for that type of car (automatic or manual) for at least three years and still hold one.

Fitting car locks is wise, but this is ridiculous!

(iii) Using the lights and horn

Questions

1 Why is it important that a vehicle's headlamps should be correctly adjusted?

2 When must I switch on my lights?

3 When must I use headlamps?

4 What lights should I use at night in a built-up area?

5 When should I dip my headlamps?

6 What action should I take if I am dazzled by the lights of an approaching car?

Answers *(Using the lights and horn)*

1 Badly adjusted headlamps can dazzle other road users — and cause accidents.

2 At lighting-up time. This varies according to the season.

3 There are three sets of conditions when headlamps must be used. The first is at night on all roads where there is no street lighting, where the street lights do not work, or where the lights are spaced more than 185 metres (200 yd) apart. The second is when fog or

gloom restricts the visibility to a distance of less than 100 metres. This includes road tunnels and unlit underpasses. The third is when you are driving on a motorway or high-speed road at night, even if they are lit.

4 Use dipped headlamps unless the street lighting is very good.

5 The purpose of dipping your headlamps is to avoid dazzling other road users, particularly the drivers of other vehicles. Dip the headlamps as soon as you meet other vehicles, either ahead of you and heading in the same direction, or approaching you in the opposite direction.

6 Slow down or stop. Do not flash your own headlamps as a rebuke.

7 On what occasions should I flash my headlamps?

8 Are rear fog lamps useful additions to a vehicle?

9 If visibility is poor, how fast should I drive?

10 On what occasions should I sound my horn?

11 Are there any times when the sounding of horns is forbidden?

Answers

7 There is only one reason for flashing your headlamps, to show another road user that you are there. Flashing headlamps should not be used as a sign of anger, or as an invitation to drive on. There are so many ways of interpreting a flash that confusion is likely to be the main result.

8 Yes, but they should only be switched on in very foggy conditions, when visibility has been reduced to less than 100 metres. They should not be used just because it is dark or raining.

9 Make sure that you are travelling slowly enough to be able to stop well within the distance you can see ahead.

10 The rule about the sounding of horns is just the same as that concerning the flashing of headlamps. Only use the horn to show other road users that you are there. Do not sound it to show anger.

11 Yes. You must not use your horn when the car is stopped on the road, except in an emergency to advertise your presence. And you must not use your horn when driving in a built-up area between the hours of 11.30 p.m. and 7 a.m.

LANE 3

On the Road

This set of questions and answers deals with behaviour on the road. The general rules are simple. You must take the proper precautions before starting the engine and pulling out. Once on the road you must pay close attention to the traffic both in front and behind you all the time. And when you stop or park you must do so safely (and legally).

What kerb markings tell you about parking

QUESTION: These are three yellow lines which you may find marked on kerbs. What do they mean?

a) b) c)

(Answers: Page 188)

(i) Driving along

Questions

1 How should I check for traffic behind me when I want to move out from the kerb?

2 How big a gap in the traffic should there be before I can move out?

3 When and how should I give signals to other road users?

4 Whose hand signals must be obeyed on the road?

5 On which side of the road should I drive?

6 On which occasions can I move to the right-hand side of the road?

Answers *(Driving along)*

1 Always look in your mirror before you move off. As an extra precaution, turn round and actually look behind you as well.

2 This all depends on the speed and density of the traffic. On no account should you cause another road user to swerve or brake to avoid hitting you. Always leave ample room and signal if necessary.

3 Give signals as help or warnings to other drivers or pedestrians. Signal clearly and in plenty of time — and remember to turn off your indicator signals when the manoeuvre is completed. (For questions on specific signals, see page 140).

4 You must obey the signals of police officers and traffic wardens who are directing traffic, and of school-crossing patrols.

5 On the left-hand side.

6 When the road markings or road signs indicate that you must do so; when you want to overtake a slower vehicle in front of you; when you want to turn right; or when there are stationary vehicles or pedestrians in the road.

7 Am I allowed to drive on the pavement?

8 What should the driver of a large or slow-moving vehicle do to prevent the bunching of faster traffic behind him on narrow and winding roads?

9 How can I prevent myself from feeling drowsy on a long car journey?

10 What is the speed limit on all roads (except motorways) where there are street lights?

11 If a speed limit is imposed on a road, is it safe for me to drive at that speed?

12 How can I judge what is a safe speed for me to travel at?

Answers

7 No. You must not drive onto the pavement or on any footpath.

8 He should look for an opportunity to pull into the side of the road and stop in order to let the other vehicles overtake.

9 While you are driving, keep a window open or the air conditioning on so that there is a steady supply of fresh air. You are much more likely to feel sleepy in a stuffy car. If the feelings of drowsiness persist, or if you become tired, stop the car at a safe parking place and have a rest.

10 The limit is 30 mph. unless there are signs showing otherwise.

11 Not necessarily. The speed limit is the maximum speed at which you are legally allowed to travel. If the conditions are bad, you should drive at a slower speed.

12 Make sure that you will be able to stop well within the distance that you can see ahead. Drive more slowly at night, or in wet, foggy or icy road conditions.

13 How close should I drive to the vehicle in front of me?

14 What is the shortest stopping distance for a car travelling on a dry road at 20 mph?

15 What is the shortest stopping distance for a car travelling on a dry road at 40 mph?

16 What is the shortest stopping distance for a car travelling on a dry road at 70 mph?

17 How should these stopping distances be increased in wet or icy conditions?

Answers

13 Always make sure that there is enough space for you to stop safely if the vehicle in front slows down or pulls up without warning. Also leave enough space for a vehicle which is overtaking you to pull into. As a rough guide, leave a gap of one metre for each m.p.h. of your speed, so long as the road conditons are good. Leave a larger gap in wet or icy conditions.

14 The total stopping distance is 12 metres (40 ft), made up of 6 m (20 ft) thinking distance and 6 m (20 ft) braking distance.

15 A total of 36 m (120 ft), made up of 12 m (40 ft) thinking distance plus 24 m (80 ft) braking distance.

16 A total of 96 m (315 ft), made up of 21 m (70 ft) thinking distance plus 75 m (245 ft) braking distance.

17 They should be doubled at least.

18 What should I do if a vehicle overtakes and goes into the gap in front of my car?

19 What kind of vehicles should I make way for?

20 What should I do if I see a flashing amber light?

21 What should I do if I think a police car wants me to stop?

22 Are there any special points to remember when driving in residential areas?

Answers

18 Slow down and drop back so as to leave the same gap as before.

19 For all emergency vehicles — police cars, ambulances or fire appliances — with flashing lamps and sounding bells, two-tone horns or sirens. You should also allow buses to pull out from bus stops ahead of you, if it is safe to do so.

20 Drive carefully, as this warns you of a slow-moving vehicle (such as a road-gritter or tractor) or a vehicle which has broken down.

21 If the police want you to stop they will, where possible, attract your attention from behind by flashing their headlights or blue light or by sounding a siren or horn. A police officer will direct you to pull over by pointing and using the left indicator. You must pull over and stop as soon as it is safe to do so, and switch off your engine.

22 Always drive slowly in residential areas and look out for special features such as road humps and narrowing that are intended to slow you down. A 20 mph maximim speed limit may also be in force.

Why the Police may ask you to stop

(Answer: Page 189)

(ii) Keeping in lane

Questions

1 What is a single-track road?

2 What should I do if I see a vehicle coming towards me on a single-track road?

3 What should I do if a driver behind me wants to overtake on a single-track road?

4 What should I do if I meet a vehicle that is coming uphill on a single-track road?

5 Am I allowed to park in the passing places on a single-track road?

6 What is the meaning of a single broken line, with long markings and short gaps, in the middle of the road?

Answers *(Keeping in lane)*

1 A road which is not wide enough for vehicles to pass each other.

2 Look for a suitable passing place. If it is on your side of the road, then pull into it and let the other driver go by. If the passing place is on the other side of the road, i.e. the right-hand side, stop and wait opposite it so that the other driver has space to pull in.

QUESTION: Neither of these drivers is going to get through by arguing about it. One of them should have moved into the passing place on the right to let the other one through. Which one?

(Answer: Page 189)

3 Look for a passing place and use the same procedure as above.

4 Always try to give way to it whenever possible.

5 No. You will be blocking the access for someone else.

6 It is a hazard warning line. It should not be crossed unless the road is clear for a long way ahead.

7 What do unbroken double white lines in the centre of the road mean?

8 There are double white lines in the centre of the road, but the line nearest me is broken. What does this mean?

9 What is the purpose of areas marked with white diagonal stripes or white chevrons?

10 Which lane should I normally travel in?

11 What is the meaning of the red and white studs which are sometimes used to mark the roads?

Answers

7 You must not cross the lines except in special circumstances. These are: when getting in and out of premises or a side road; when ordered to cross them by a police officer or traffic warden; or when you have to move out to avoid an obstruction such as a parked car.

8 You are allowed to cross the lines to overtake the vehicle in front if you can do so safely. Make sure you can do so before reaching an unbroken white line on your side.

9 They are there to separate streams of traffic which might be a danger to each other at that point, or to protect vehicles which are turning right. You should not drive onto these areas if you can help it.

10 In the left-hand lane of any carriageway, unless you are going to overtake, turn right, or move out to pass a parked vehicle.

11 Studs are often used in conjunction with white lines. White studs mark the lanes or the centre of the road. Red studs mark the left-hand side of a carriageway. Amber studs mark the edge of the central reservations of dual carriagways. Green studs are used to mark the boundaries of lay-bys or side roads.

12 What precautions should I take when moving into another lane?

13 What should I look for to guide me at junctions?

14 In a traffic hold-up, should I try to overtake vehicles which are waiting in front of me by moving into another lane?

15 If there are three lanes on a single, undivided road, what should the middle lane be used for?

16 If a single-carriageway road has four lanes or more, am I allowed to use the lanes on the right-hand side?

17 On a two-lane dual carriageway what is the purpose of the right-hand lane?

Answers

12 First of all, look in your mirrors. Judge whether there is enough time and space for you to move out without causing other traffic to swerve or slow down. When it is safe, signal before moving over.

13 Look for lane indication arrows marked on the road or on signs.

14 No. Cutting into other lanes and trying to jump the queue will only anger other road users. It could also cause an accident, and, in the end, it will probably gain you only a small advantage.

15 The middle lane should be used for overtaking or for turning right, and for no other reason. Do not under any circumstances use the right-hand lane, and do not linger in the middle lane.

16 No. You must not use them unless there are signs or markings to show that you may do so.

17 It is for overtaking or turning right only.

18 On a three-lane dual carriageway, am I allowed to stay in the middle lane?

19 What is the right-hand lane of a three-lane dual carriageway used for?

20 Which lane should I use in a one-way street with three lanes?

21 What is particularly hazardous about three-lane one-way streets?

22 Am I allowed to drive in a bus lane?

23 Am I allowed to drive in a cycle lane?

24 What is a 'crawler' lane?

Answers

18 You may stay in the middle lane only as long as there is slower-moving traffic in the left-hand lane. You should move back to the left-hand lane as soon as you can safely do so.

19 For overtaking or for turning right. After overtaking you should move back into the middle lane as soon as you can, and then back into the left-hand lane. Make sure that you do not cause other traffic to swerve or change speed.

20 If you are going to the left, use the left-hand lane; if to the right, use the right-hand lane; if you are going straight on, use the centre lane. However, watch the road markings closely, because they may indicate otherwise. Never change lanes suddenly.

21 Other traffic may be passing on both sides of you.

22 Not while the lane is in operation. Bus lanes are indicated by road markings and signs, which show the vehicles, such as taxis, which may use them. The signs will also show whether the bus lane is in operation 24 hours a day or only at particular times of the day. When the lane is not in operation it may be used by any vehicle.

23 No. If the cycle lane is marked by an unbroken white line, then it is an offence to drive or park in it. If the lane is marked by a broken white line, drivers should avoid driving in it if at all possible.

24 On some hills an extra uphill 'crawler' lane may be provided. Use this lane if you are driving a slow-moving vehicle, or if there are vehicles behind you wishing to overtake.

Motoring Firsts • 2

The first motor car to be stolen, in Paris in 1896, was a Peugeot, taken from the manufacturers by the owner's mechanic. Ironically, the same year, five months later, the first motor insurance was introduced in Britain by the General Accident Company.

The first motor coach excursion ran in the Blackpool area in 1897, and the first motor coach tour took place the following year. It lasted six days and toured France from Paris to Aix-les-Bains in eastern France.

(iii) Parking

Questions

1 When parking, what are the main things to keep in mind?

2 Where is it illegal to stop or park?

3 The Highway Code lists 10 examples of places where it would be dangerous to park or where parking would inconvenience pedestrians and other road users and in doing so would likely be committing an offence. What are they?

Answers *(Parking)*

1 (a) Wherever possible, pull off the road on to an area provided for parking.

(b) If you have to park on the road, stop as close to the side as you can.

(c) Leave plenty of room when parking next to or behind a vehicle displaying a disabled person's badge.

(d) Before you or your passenger open the car door make sure it will not hit anyone passing on the road or force them to swerve.

(e) It is safer for both you and your passengers to get out of the car on the side of the road next to the kerb.

(f) You must switch off the engine and headlights. Also ensure that the handbrake is on firmly and the vehicle is locked.

2 (a) On a motorway (except, in an emergency, on the hard shoulder).

(b) On a pedestrian crossing, including the area near a Zebra crossing marked by zigzag lines and the area near a Pelican crossing marked by rows of studs. (You must, of course, stop to allow pedestrians to cross.)

(c) On a road marked by signs as a Clearway.

(d) On a road marked as an Urban Clearway during the period shown on the signs (except for stopping briefly in order to set down passengers).

(e) On any section of road marked with double white lines even if one of the lines is broken (except to set down passengers or to load and unload goods).

(f) On a bus, tram or cycle lane during the periods that they are in operation (except, where permitted, to load or unload goods).

(g) On the side of a road marked with yellow lines at the edge (or red lines in the case of specially designated 'red routes') during the restricted times shown on the plates (except to set down passengers or to load and unload goods).

(h) In a parking space reserved for specific users such as Orange Badge holders or residents, unless entitled to do so.

(i) You must not park on the road in such a way that your vehicle or trailer is likely to cause danger to other road users or an unnecessary obstruction.

QUESTION: This car is parked illegally. Do you know why, specifically?

(Answer: Page 189)

3 (a) On a footpath, pavement or cycle track.

(b) Near a school entrance.

(c) At or near a bus stop or taxi rank.

(d) On the approach to a level crossing.

(e) Within 10 metres of a junction, except in an authorised parking place.

(f) On the brow of a hill or hump bridge.

(g) Opposite a traffic island or another parked vehicle (if doing so would cause an obstruction).

(h) Where you would force other traffic to enter a tram lane.

(i) Where the kerb has been lowered to help wheelchair users.

(j) In front of the entrance to a property.

4 What should I remember about parking in fog?

5 What should a driver of a goods vehicle remember about loading and unloading?

6 What are the hazard warning lights on a vehicle?

7 When should the hazard warning lights be switched on?

8 Am I allowed to park a vehicle on a road at night?

9 Which vehicles are legally allowed to park on a road at night?

Answers

4 It is especially dangerous to park on the road in fog, but if it is unavoidable leave your sidelights on.

5 (a) Restrictions on loading and unloading are shown by yellow markings on the kerb. Loading and unloading may be permitted when parking is otherwise restricted.

(b) Goods vehicles with a maximum laden weight of over 7.5 tonnes (including any trailer) *must not* be parked on a central reservation without police permission or on a verge or footway, except where this is essential for loading or unloading (in which case the vehicle must not be left unattended).

6 The four direction indicators flashing simultaneously; they are operated by a single switch.

7 Only when the vehicle is stationary. The hazard lights are switched on to show that the vehicle is blocking the flow of traffic, because it has broken down or is loading or unloading. The lights should not be used when the vehicle is in motion, or while stopping for a trivial reason.

8 Not if you can possibly avoid it. If you have to park on a road at night, you must by law switch on your lights. However, certain vehicles are exempt.

9 Cars and goods vehicles not more than 1,525 kg unladen, invalid carriages and motorcycles may be parked without lights on a road with a speed limit of 30 mph. But they must fulfil certain conditions: they must be parked at least 10 metres from a junction, close to the kerb and facing the same way as the flow of traffic; or they must be parked in a recognised parking place.

Other vehicles and trailers, and all vehicles with projecting loads *must not* be left on a road at night without lights.

Cross Quiz

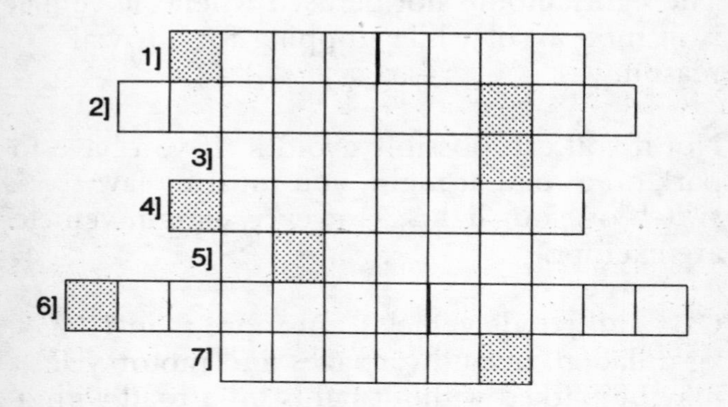

ALL THE CLUES ARE FOR ACROSS ANSWERS ONLY.

1] Not a main road
2] You look through it when driving
3] Your car radio uses one
4] Extra lights for driving in fog
5] They keep [2] above clear in wet conditions
6] Electro mechanical device for getting your car going
7] It makes cars shine.

THE LETTERS IN THE SHADED SQUARES SPELL SOMETHING NO DRIVER SHOULD EVER BE.

(Answers: Page 189)

LANE 4

Manoeuvres

Getting safely onto the road is only the first step. Before you go very far you are likely to have to get off it again, either to turn into another road or to make your way round a roundabout. The following set of questions deals with the many sorts of manoeuvres you will have to make — turning, overtaking, reversing and negotiating road junctions.

What is wrong in these pictures?

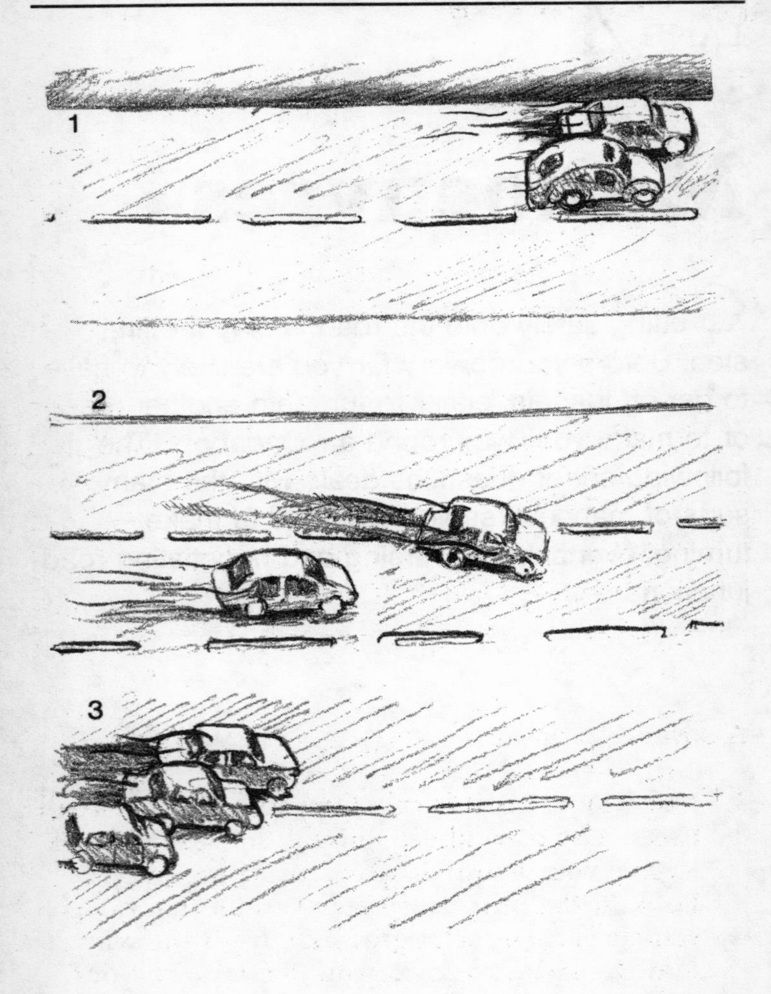

(Answers: Page 189)

(i) Overtaking

Questions

1 What must I check before I overtake another vehicle?

2 What extra precaution should a pedal cyclist or motorcyclist take?

3 What kind of conditions make it difficult to judge speed and distance?

4 Why should I keep a particular look-out for cyclists and motorcyclists when I am overtaking?

5 What danger must I look out for when preparing to overtake on a fast road?

Answers *(Overtaking)*

1 Make sure that it is safe for you and other road users. The road ahead should be clear enough to give you plenty of space to manoeuvre. Look in the mirror to make certain that no vehicle is approaching quickly from the rear: it may decide to overtake you. Signal with your indicators well before you start to move out.

2 Look behind you and to your off-side.

3 Dusk, darkness and mist or fog.

4 Pedal cycles and motorbikes are much less easy to spot than other, wider vehicles. Drivers of all vehicles (especially long vehicles or ones towing trailers) should leave plenty of room for cyclists. Cyclists have just the same rights on the road as any other driver, but are much more vulnerable.

5 Vehicles behind you will be approaching at speed, often much faster than you think. You should ensure that the road behind you is clear to a greater distance than on slower roads.

What's the problem here?

(Answer: Page 190)

6 Once I have started to overtake, what procedure should I follow?

7 What precautions should I take when overtaking motorcycles, pedal cycles or horse riders?

8 On what occasions am I allowed to overtake another vehicle on its left-hand side?

9 When the traffic is moving slowly in separate lanes, am I permitted to change lanes to the left in order to overtake?

10 What should motorcyclists watch out for when overtaking traffic queues?

Answers

6 Move as quickly as possible past the vehicle you are overtaking. Do not drive too close to it, but leave plenty of room. When you are safely past, move back to the left-hand side of the road. Once again, leave plenty of room and do not cut in sharply.

7 Always give them a wide berth. Cyclists and horse riders should be given as much room as you would a car. In windy conditions, cyclists may be blown off a straight course. Where the road surface is poor they may have to swerve to avoid obstructions. Horses may be nervous. Finally, do not overtake cyclists or horse riders just before you are to take a left turn.

8 There are four occasions:

(a) When the driver in front has signalled that he means to turn right, and there is enough room for you to pass by on the left-hand side without obstructing others or entering a bus or tram lane in operation.

(b) When you wish to turn left at a junction.

(c) When the traffic is moving slowly in separate lanes and the vehicles in the right-hand lane are moving more slowly than you are.

(d) In one-way streets where vehicles can pass on either side.

9 No. You may only move into a lane to your left if you intend to turn left or to park.

10 They should look out for pedestrians crossing between vehicles.

11 What should I *not* do when another vehicle is overtaking mine?

12 If I am driving on a two-lane road, what rule should I follow when moving out to pass a parked vehicle or other obstruction on the left-hand side of the road?

13 When must I not, by law, overtake another vehicle?

14 In what other conditions should I not overtake another vehicle?

Answers

11 You should not increase your speed. This is a stupid and dangerous way to behave. If necesssary, you should be prepared to slow down and let the overtaking vehicle come by more quickly.

12 You should give way to any vehicles approaching you from the opposite direction.

13 You must not overtake if:

(a) You have to cross or straddle double white lines when the solid line is the nearer to you.

(b) You are within the zigzag area on the approach to a Zebra crossing.

(c) You are within a 'No Overtaking' zone marked with a sign.

14 You should not overtake if:

(a) You cannot see far enough ahead to be certain that it is safe to do so (near, for example, a bend, the brow of a hill or a hump-backed bridge).

(b) You might obstruct or annoy other road users (for example, at a road junction, a level crossing, a narrow stretch of road or an area marked with diagonal stripes or chevrons; when approaching a school crossing patrol; where you would have to enter a bus, tram or cycle lane; between a bus or tram when it is at a stop; when there is a possibility you would cause other vehicles to swerve or slow down).

(c) There is a possibility that you would cause other vehicles to swerve or slow down.

(d) You are in any sort of doubt at all.

Did you know that • 1

When driving tests were first introduced in Britain in March 1935, they were voluntary? This did not last long, though, driving tests became compulsory three months later, on 1st June 1935 and so did the carrying of L-plates for learning-drivers.

(ii) Road junctions

Questions

1 What precautions should I take when approaching a road junction?

2 What particular dangers should I look out for at road junctions?

3 I am waiting to come out of a road junction, and an approaching vehicle on the right is indicating that it is going to turn left. Should I pull out?

4 What is the meaning of double broken white lines across the road at a junction (which may also be marked by a 'Give Way' sign or an inverted triangle?

5 At what kind of junction am I obliged to stop before proceeding?

6 How should I cross or turn right into a dual carriageway?

Answers *(Road junctions)*

1 You should approach with great care, keeping a close eye on your speed and road position. Drive on past the junction only when there is room for you to do so without impeding other road users.

What is a safe gap?

(Answer: Page 190)

2 Beware of long vehicles which may be turning either to the right or to the left at the junction ahead. They may be using the entire width of the road in order to make the turn. Look out also for motorcyclists, pedal cyclists and pedestrians.

3 Do not pull out until you are certain that the vehicle really is going to turn left. The driver may have left his indicator on by mistake and could carry on along the major road.

4 You must give way to vehicles travelling on the major road.

5 At junctions marked with a 'Stop' sign and a solid white line across your approach. Always wait until there is a safe gap in the traffic before moving out from the junction.

6 A dual carriageway should be treated as two separate roads. When it is safe to do so, move into the central reservation and wait until there is a suitable gap in the traffic. Then carry on across or into the further lane.

7 What should I do if the central reservation is too narrow for the length of my vehicle?

8 How is a box junction marked?

9 What are the rules governing traffic at a box junction?

10 When all traffic at a junction is being help up by a police officer or traffic warden, may I use the filter lanes to left or right?

11 At a junction controlled by traffic lights, may I move forward when the red and amber lights are showing together?

Answers

7 Wait in the side road until there is a suitable break in the traffic coming from both directions. Then cross in one movement.

8 By criss-cross yellow lines painted on the road.

9 Drivers may not enter the box if their exit road is not clear and is blocked by other vehicles. However, they can enter the box and wait if they are turning right and are prevented from doing so by oncoming traffic or vehicles on the other side also wanting to turn right.

10 You may, but only when the officer or warden signals you to do so.

11 No, under no circumstances.

12 Can I automatically move forward when the traffic lights are green?

13 What are the rules governing filter lanes at traffic lights?

14 Where must vehicles wait when they are required to stop at junctions controlled by traffic lights?

15 What type of road users are particularly vulnerable at junctions?

16 What should I do if traffic lights are not working?

Answers

12 Not necessarily. First check to make sure that there is room for you to clear the junction in complete safety.

13 You should not get into the filter lane (ie the lane indicated by a green arrow filter signal on the lights) unless you want to go in the direction shown by the arrow. Give other road users plenty of room to move into the filter lane.

14 Behind the solid white line marked 'Stop' across the approach.

15 Pedal cyclists, motorcyclists and pedestrians. Always look out for them before you turn and give them room.

16 You should proceed with great caution.

Motoring Firsts • 3

The first set of traffic lights predated the motor by some seventeen years. They were set up in Parliament Square, London, in December 1868 to control horse-drawn traffic. The lights stood on top of a 22-foot cast iron pillar. A gas-powered revolving lantern, operated by a lever, was used to change the lights. They had only two signals: red for STOP and green for CAUTION.

When the first successful petrol-driven car was built by the German Karl Benz in 1885, many people believed that an engine powered in this way was bound to explode. Because of this, another German, Gottlieb Daimler, provided his engine with a mass of wires and insulators when he launched the first petrol-driven motorboat in 1886. This was to allay people's fears by giving the impression that the motorboat was powered by electricity.

(iii) Turning left and right

Questions

1. Before turning right, when and why should I use my mirrors?
2. What sequence of actions should I then follow to turn right?
3. Should I give way to pedestrians crossing the road into which I am turning?
4. If I am turning right at a junction and there is an oncoming vehicle also turning right, how should I position the vehicle?

Answers *(Turning right and left)*

1. Look in your mirrors well before you intend to turn right. This will show you the position and movement of the traffic behind you.

2. When you are sure that it is safe, indicate that you are going to turn right. Take up a position just to the left of the middle of the road or in the space marked for right-turning traffic. If possible, leave room for vehicles behind you to pass by on your left-hand side. When there is a safe gap in the oncoming traffic, complete the turn. Be careful not to cut the corner.

3 Yes. You should also look out for oncoming cyclists and motorcyclists.

4 Keep the oncoming vehicle on your right-hand side and pass behind it, off side to off side. Look out for other traffic on the carriageway you intend to cross before moving past. If it is impossible to pass behind the other vehicle (either because there is not enough space, or becuse road markings indicate otherwise) then you will have to pass near side to near side (ie in front of it). In this case be especially careful in watching for oncoming traffic, as your view of it may be obscured by the other vehicle.

5 What is the procedure for turning right from a dual carriageway?

6 What should be my sequence of actions when turning left?

7 Where should I position my vehicle on the road when turning left?

8 What are the main dangers posed to cyclists, motorcyclists and pedestrians when I turn left?

9 What should I look out for when turning left across a bus, tram or cycle lane?

Answers

5 Move into the opening in the central reservation. Wait there until there is a safe gap in the oncoming traffic in the other carriageway. Then move across.

6 Look in your mirrors well before you intend to turn left. Give a left turn signal. Then turn.

7 Before and after the turn keep as close as you safely can to the left-hand side of the road.

8 You should not overtake a cyclist or motorcyclist immediately before turning left. Make sure that one is not approaching you from behind on your left. Give way to pedestrians crossing the road into which you are turning.

9 Buses, trams and bicycles, of course! And give way to them.

(iv) Roundabouts

Questions

1 What must I do when approaching a roundabout?

2 What is the rule about giving way at a roundabout?

3 How should I turn left at a roundabout?

4 If there are two lanes at the entrance to a roundabout and I wish to go straight ahead, which lane should I drive in?

5 If there are two lanes at the entrance to a roundabout and I wish to go right, or turn full circle, which lane should I drive in?

6 Which lane should I take if there are more than two lanes at the entrance to a roundabout?

Answers *(Roundabouts)*

1 Decide as early as possible which exit you need to take and get into the correct lane. Reduce your speed and check to see what vehicles are already on it. Look out particularly for pedal cyclists or motorcyclists ahead or to the side. At some junctions there may be more than one

roundabout. At each one use the normal rules for roundabouts.

2 Give way to traffic coming from your right unless the road markings show otherwise. If there is nothing coming from the right, keep moving.

3 Signal left and approach in the left-hand lane; keep to the left of the roundabout and continue signalling left as you turn left.

4 Approach in the left-hand lane, and follow it round the roundabout. If there is an obstruction in the left-hand lane, then approach in the right-hand lane and follow that round the roundabout. If the roundabout is clear of traffic, take the most convenient lane through.

5 Approach in the right-hand lane, and follow it round the roundabout.

6 Use the clearest convenient lane both approaching and driving through the roundabout, depending on which exit you wish to take.

7 What points should I look out for when driving round a roundabout?

8 How should I signal when turning left at a roundabout?

9 How should I signal when going straight ahead?

10 How should I signal when turning right?

11 What kind of vehicles may take a misleading course over a roundabout?

12 Do the same rules apply to mini-roundabouts?

Answers

7 Pay close attention to other vehicles who may have to cross in front of you in order to reach the next exit. Be particularly watchful of pedal cyclists, motorcyclists, and horse riders.

8 Use the left turn indicator on approach and through the roundabout.

9 Use the left turn indicator when passing the exit immediately before the one which you will be taking.

10 Use the right turn indicator as you approach, and keep this going until you pass the exit immediately before the one which you will be taking. Then change to the left turn indicator.

11 Long vehicles. They may have to approach and cross the motorway at a different angle from shorter vehicles.

12 Yes and watch out for long vehicles which may have to cross the centre of a mini-roundabout.

"We thought this was a good place to stop for lunch!"

Never halt on a hard shoulder unless in an emergency.

(v) Reversing

Questions

1 What must I check for before I reverse a vehicle?

2 What should I do if I cannot see clearly behind?

3 Am I allowed to reverse out of a side road into a main road?

Answers *(Reversing)*

1 Make sure that there are no obstructions or pedestrians (especially children) behind you. Remember that there will be a part of the road behind you which you cannot see. This is called the 'blind area'.

2 Get someone to guide you while you reverse.

3 Certainly not. It is a very dangerous manoeuvre. As a general principle, you should not reverse your vehicle for a longer distance than is absolutely necessary.

Roundabout Word Puzzle

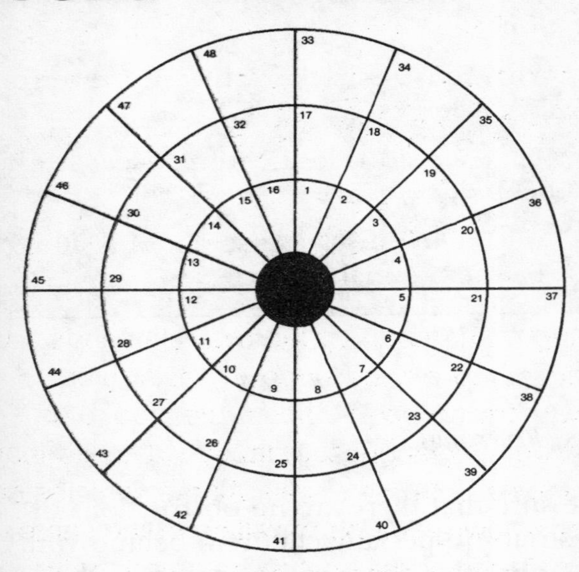

In this puzzle, from the second word on, each successive word starts with the letter that ends the preceding word.

Clues:

1-5 70 mph is the maximum allowed in Britain.
5-16 Describes the slowing down of a vehicle.

17-29 You'll reach a dead end if you drive down a road bearing this sign.
29-32 When driving with full headlights at night, a good driver ---- his/her headlights when another car approaches coming the other way.

33-36 You drive along this road on the approach to a motorway.
36-48 On most of them, you can stay for two hours only.

(Answers: Page 190)

LANE 5

Motorway Driving

Driving on an ordinary highway demands a lot of attention and a knowledge of basic rules. Motorway driving needs those rules too, plus some extra special ones. In many ways, life may seem simpler on a motorway. All the vehicles on your carriageway will be travelling in the same direction. There should be no hold-ups, traffic lights or roundabouts to worry about. But these conditions can unleash two major dangers: firstly, it is all too easy to travel too fast; and secondly, it is easy to lose your concentration.

Motorway Quiz 1

QUESTION: Which of these signs will you NOT see on a motorway?

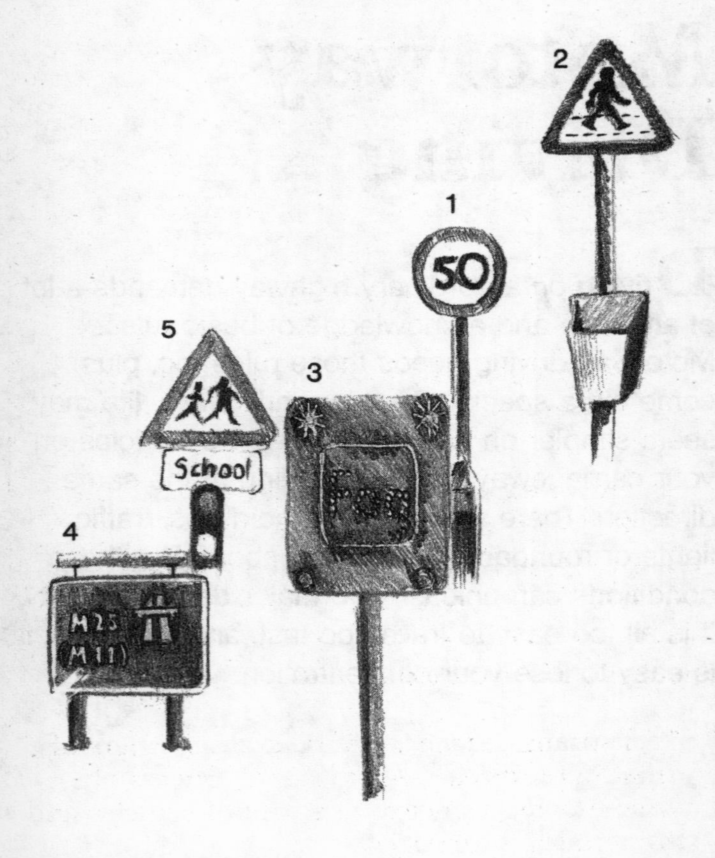

(Answer: Page 190)

(i) Joining a motorway

Questions

1 What kinds of road users are prohibited from using a motorway?

2 Is a driver allowed to pick up or set down a passenger on the slip road of a motorway?

3 What major difference in road conditions will I find when I drive on a motorway?

4 Should I check my vehicle before driving on a motorway?

5 Are slip roads and link roads between motorways always straight so that they can be taken at speed?

6 From which side of a motorway will the slip road normally approach?

Answers *(Joining a motorway)*

1 Pedestrians, learner drivers, cyclists, riders of small motorcycles, horse riders, drivers of slow-moving vehicles, agricultural vehicles and some invalid carriages.

2 No. It is an offence to do this on *any* part of a motorway.

3 The traffic will be travelling faster than on ordinary roads. You will have less time to sum up situations and will need to think more quickly. Using your mirrors and concentrating all the time are very important.

4 This is advisable. Make sure that it is in a fit condition to be driven at speed over a long distance. Check the pressure of the tyres and the levels of petrol, oil and water. Ensure that you have at least enough petrol to get you to the next service area. If you are carrying or towing an extra load, make sure that it is securely fixed.

5 No. Some may have sharp bends for which you must reduce speed.

6 From the left (unless perhaps you are joining at the start of the motorway).

7 What procedure should I follow when joining a motorway?

8 Should I immediately move into the overtaking lanes?

9 Am I allowed to turn round on a motorway?

10 What should I do if I miss my turn-off point or take the wrong route?

11 How fast should I drive on a motorway?

12 How can I help to prevent myself feeling sleepy on a long journey?

Answers

7 Watch for a safe gap in the traffic in the left-hand lane of the motorway as you approach. Then adjust your speed in the extra (acceleration) lane so that when you join the motorway you are already travelling at the same speed. If there is not a suitable gap in the traffic, wait in the acceleration lane until it is safe to move out onto the motorway. Always give way to traffic already on the motorway.

8 No. It is a good idea to stay for a while in the left-hand lane until you have got used to the speed of the traffic. At some junctions the slip-road continues as an extra lane on the motorway. Where signs indicate that this will happen, stay in that lane until it becomes part of the motorway.

9 Certainly not. You must not turn in the road, or reverse, or cross the central reservation, or drive against the traffic.

10 You must stay on the motorway until you reach the next exit. Never try to turn or reverse (see previous question).

11 In good weather conditions, you should cruise at a steady speed to suit the limits of your vehicle. You must not, of course, exceed the speed limit (see page 185). If the road is wet or icy, or there is fog or rain about, reduce your speed accordingly.

12 Make sure that there is plenty of fresh air in your vehicle: keep windows partly opened and switch on air conditioning. If this does not work, turn off at the next exit or service area, park the car and walk around for a few minutes. If in doubt, stay off the motorway.

Motorway Quiz 2

QUESTION 1: "Motorway"is the word used in Britain for this kind of fast road. Here are words used in other countries. Can you match up each word with the countries listed?

(1) AUTOBAHN	(A) SWEDEN
(2) AUTOSTRADA	(B) FRANCE
(3) FREEWAY	(C) GERMANY
(4) AUTOROUTE	(D) SPAIN
(5) AUTOPISTA	(E) ITALY
(6) MOTORVAG	(F) UNITED STATES

QUESTION 2: Motorways are not the only fast roads. Dual carriageways are, too. However, what are the differences between them?

(Answers: Page 190)

(ii) Motorway manoeuvres

Questions

1 On a two-lane carriageway, which lane should I drive in?

2 Which lanes should I use on a three-lane carriageway?

3 What are the rules for using the right-hand overtaking lane?

4 What do the coloured studs at the edge of the carriageways mean?

5 What kind of vehicles are not allowed to use the right-hand lane?

Answers *(Motorway manoeuvres)*

1 In the left-hand lane unless you are overtaking another vehicle.

2 Keep to the left-hand lane whenever possible. use the middle lane for overtaking slower traffic. You may stay in the middle lane when there are several slower vehicles in the left-hand lane, but you should return to the left-hand lane as soon as you have passed them. The right-hand lane is not a 'fast lane'. It is an overtaking lane.

3 Use it for overtaking only. When you have passed the vehicles in the middle lane, move back into the middle lane. Then move into the left-hand lane without cutting in sharply. Do not linger in the middle or right-hand lanes unnecessarily.

4 Amber-coloured studs mark the right-hand edge of the carriageway. Red studs mark the left-hand edge. Green studs mark the separation between the acceleration and deceleration lanes and the motorway itself.

5 A goods vehicle with an operating weight of more than 7.5 tonnes, or any vehicle pulling a trailer, or a bus or coach longer than 12 metres. These must not use the right-hand lane except in emergencies.

6 On a motorway, am I allowed to overtake vehicles on their left-hand sides?

7 Am I allowed to use the hard shoulder for overtaking?

8 What is the procedure for overtaking on a motorway?

9 What sort of road conditions should make me especially careful?

10 How should I proceed once I have overtaken the vehicle in front?

Answers

6 No. You should never overtake on the left, except in special circumstances. These are: when the traffic is moving slowly in queues and the queue on your right is moving more slowly than you are. Never move to a lane on your left simply to overtake.

7 No, not under any circumstances.

8 First of all make sure that the lane into which you will be moving is clear for a safe distance behind and ahead. Remember that traffic will be approaching from behind more quickly than you think. Use your mirrors. Signal before you move out.

9 At dusk, in the dark and in fog or mist. In these conditions it is more difficult to judge speed and distance correctly.

10 As soon as you can, move back to the left-hand lane. If this is occupied, wait in the middle lane until there is a safe gap. Do not cut in on the vehicle you have just passed.

(iii) Breakdowns and other emergencies

Questions

1. What should I do if my vehicle breaks down?
2. How will I know which way to walk for the nearest emergency telephone?
3. Should I leave the passengers in my vehicle?
4. What should I do if I cannot move my vehicle onto the hard shoulder?
5. What are the rules for re-joining the motorway from the hard shoulder?

Answers *(Breakdowns and other emergencies)*

1 Get it off the motorway and onto the hard shoulder as quickly as possible. Park it as far to the left as you can. Switch on your hazard warning lights and at night switch on your side lights as well. Your main danger is from passing traffic, so do not open the doors nearest to the carriageway and do not stand at the rear of the vehicle or on its right-hand side. Summon help by using the emergency telephones on your side of the motorway

(never cross the motorway to use the telephones on the other side). The emergency phones are free and connect you directly to the police. Give full details — tell them if you are a woman travelling alone — then return to your vehicle. If you feel at risk return to your vehicle by a left-hand door and lock all doors.

2 Look for a marker post at the back of the hard shoulder. This will bear an arrow pointing in the direction of the nearest telephone.

3 It is better not to. Remember that there is always the danger of a vehicle swerving accidentally onto the hard shoulder and colliding with your car. If the passengers do get out they should not stand behind the vehicle

Motorway Quiz 3

QUESTION 1: What is the purpose of the fast lane on the motorway?

QUESTION 2: On which side of the motorway will you find the exit slip road which leads you back to the "ordinary" road system?

(Answers: Page 191)

or walk along the hard shoulder. Children should be kept under strict control and animals should be left inside the vehicle.

4 Switch on your hazard warning lights; leave your vehicle only if you are sure you can safely get clear of the carriageway; if in doubt, remain in your vehicle wearing your seat belt until the emergency services arrive. *Do not* attempt to place a warning triangle on the carriageway.

5 Do not drive straight onto the motorway. Build up your speed first on the hard shoulder. Watch for a safe gap in the traffic before moving out.

6 What should I do if something falls from my vehicle or another vehicle and might cause danger to traffic?

7 What signals are shown on the motorway if the conditions are dangerous?

8 How far apart are the emergency signals placed?

9 To which lanes do the signals apply?

10 What must I do if the red lights on the signal begin to flash?

11 What do the flashing amber signals at the entrances to some motorways mean?

Answers

6 Stop your vehicle on the hard shoulder and use the emergency telephone to notify the police. They will arrange for the object to be picked up. Do not try to retrieve it yourself.

7 Amber lights flash and the central panel of the signal indicates a temporary maximum speed. It may also show that certain lanes are closed. When the danger has been passed the panel of the next signal will show the end of the restriction. There will no flashing lights. In normal conditions the signals will be blank.

8 They are not more than two miles apart.

9 If they are on the central reservation, the lights apply to all lanes. On some very busy motorways the signals are overhead, one applying to each lane.

10 If the red lights are above your lane, you must not pass beyond the signal in that lane. If the red lights are flashing on a slip road, you must not drive on it at all.

11 These signals are a warning of danger when they are flashing. The dangers include accidents, fog or risk of skidding. When the lights are flashing, restrict your speed to below 30 mph until you are sure it is safe to go faster.

(iv) Stopping, parking and leaving the motorway

Questions

1. On what occasions am I permitted to stop on a motorway?
2. What part of the motorway am I allowed to park on?
3. Am I allowed to walk on any part of the motorway?

Answers *(Stopping, parking and leaving the motorway)*

1 You must not stop on a motorway except in the following circumstances:

(a) In an emergency (ie to prevent an accident).

(b) When your vehicle breaks down.

(c) When you are signalled to stop by a police officer, an emergency traffic signal or flashing red lights.

2 You must not park on any part of the motorway itself. That includes the carriageway, the slip roads, the hard shoulder (except in an emergency) and the central reservation. You are only permitted to park at a service area. You *must not* pick up or set down anyone on a slip-road or any other part of the motorway.

3 You must never walk on the carriageway itself. Nor should you walk on the hard shoulder except in an emergency. Ensure that you keep animals and children off the carriageway and hard shoulder.

4 What rules should I observe when approaching roadworks?

5 How will I normally leave a motorway?

6 What procedure should I follow when leaving the motorway?

7 What must I be especially careful of when leaving a motorway

Answers

4 Read the signs and indications carefully, and take note of the altered speed limits. You will probably have to change lanes, so look in your mirrors, get into the appropriate lane and adjust your speed to that of the other traffic in the lane. Keep a safe distance behind the vehicle in front.

Do not switch lanes to overtake queuing traffic or drive through an area marked off by traffic cones. Watch out for traffic entering or leaving the works area, but do not be distracted by what is going on there. You must not exceed any temporary maximum speed limit.

5 On a slip road to the left-hand side.

6 Look out for the signs which warn that your turn-off is getting near. Move into the left-hand lane (if you are not already in it) well before you reach the turn-off. Stay in that lane and do not be tempted to move out and overtake shortly before reaching the turn-off point. Indicate that you will be turning left in good time. Reduce your speed if necessary to match that of vehicles in front of you in your lane. Move into the deceleration lane as soon as is safely possible and slow down before you join the slip road. At some junctions a lane may lead directly off the motorway.

7 Remember that you will have become accustomed to a higher speed on the motorway than should be used on an ordinary highway. Look carefully at your speedometer and adjust your speed to suit the new conditions — otherwise you may be driving faster than you realise.

Did you know that? • 1

There were only two motorists in Britain a century ago, in 1894? Ten years later, there were about 800 cars in the country and in 1912, some 9,000 cars and 8,000 motor taxis in London alone.

Elephants have been used for transport in India for centuries, and still are. Like all other road traffic they have to halt at traffic lights in big cities, like Bombay or New Dehli.

LANE 6

Stay Safe

There is danger nearby at any time on the road. But some situations are more dangerous than others. Fog, ice, crowded streets, pedestrian crossings, level crossings — all demand special care. The questions in this section are about hazardous conditions, how to drive in them and what to do if an accident occurs.

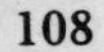

(i) Driving in fog

Questions

1 What is the first thing I should do if I drive into an area of fog?

2 Is it a good idea to be guided by the tail lights of the car in front?

3 Can foggy conditions impair my judgement of speed?

4 What lights should I use?

Answers *(Driving in fog)*

1 First of all check in your mirrors to see what is behind. If it is safe to do so, slow down to a sensible speed, depending on the conditions. Remember to keep a safe distance from the car in front. Make sure that you are able to pull up within your range of vision.

2 No. Hanging onto the tail lights ahead of you can give you a false sense of security and make you feel that you do not need to concentrate so hard. Apart from this, the vehicle in front may be travelling dangerously fast for the conditions.

3 Yes. Keep a close eye on your speedometer. You may be going much faster than you think. Do not accelerate to get away from a car close behind you. When slowing down, use the brakes so that your brake lights warn drivers behind you.

4 Make sure that you can see and be seen. If you have front fog lamps, use them. If not, use dipped headlamps. Rear fog lamps should only be used when visibility is very bad.

5 What should I check to make sure that I can see as clearly as possible?

6 What should I beware of when the fog seems to be clearing?

7 If I am setting out to drive in fog, should I make any allowances over journey times?

Answers

5 Check your windscreen and windows and use the windscreen wipers and demister to clear them. When you have an opportunity, get out and clean the outside of the screen and windows, plus the lights and reflectors.

6 You may only have hit a clear patch. Remember that fog can drift quickly and may

have clear and dense patches. You may suddenly find yourself in thick fog again, so do not relax.

7 Yes. You should allow yourself more time.

Did you know that? • 2

Motoring and motorists were news at the turn of the century. In July 1902, AUTOCAR magazine reported that a member of its staff, while in the West End of London, counted twenty-three cars within forty-five minutes.

In 1897, when Daimler cars made a publicity run from Land's End to John O'Groats, leaflets advertising their virtues were handed out along the way. This is what they read:

"No, it can't explode — there's no boiler. It can be started in two minutes. It can be stopped in ten feet when travelling at full speed (12 mph). It costs less than three-quarters of a penny a mile to run. The car can carry five people. It can get up any ordinary hill. It was built by the Daimler Motor Company of Coventry and cost £370."

(ii) Winter driving

Questions

1 How do I prepare my vehicle for winter driving?

2 Is it safe to drive in icy conditions if roads have been gritted?

3 Should I drive in snow?

Answers *(Winter Driving)*

1 Make sure that your battery is well-maintained and that there are appropriate anti-freeze agents in the radiator and windscreen-washer bottle.

2 Yes, but great care must still be taken as some roads may be slippery, and surface conditions can change abruptly in freezing or near freezing conditions.

Take care when overtaking gritting vehicles, particularly if you are riding a motorcycle.

3 No, unless your journey is essential. If it is, drive slowly and keep in as high a gear as possible to avoid wheel spin. Avoid harsh acceleration, steering and braking. Use headlights when visibility is reduced by falling

snow. Also watch out for snow ploughs as they may throw snow out from either side. Do not overtake them unless the lane you intend to use has been cleared of snow.

Castle Crossword

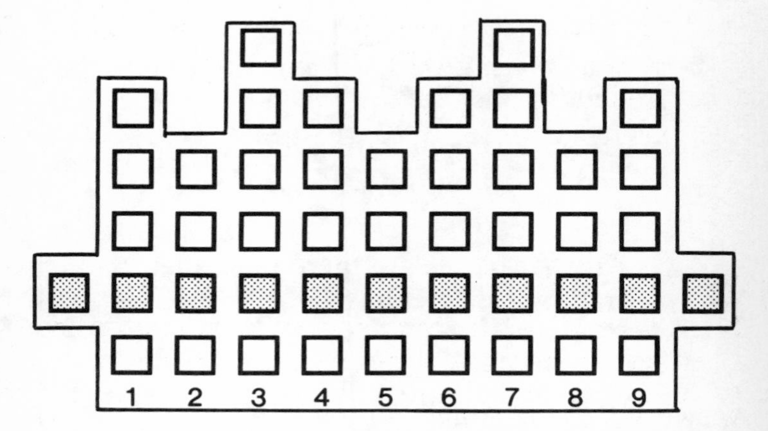

THE LETTERS IN THE SHADED HORIZONTAL SQUARES SPELL SOMETHING EVERY DRIVER SHOULD KNOW.

All the clues have answers which read *downwards* only.

1 It, too, uses a level crossing.
2 They hinder visibility, especially in winter.
3 A round road sign indicates this limit, banning some large vehicles from driving under bridges.
4 The streets turn white in these conditions.
5 The back end of your car.
6 There are car parks for long ones and for short ones.
7 There are no through roads down cul -- ----.
8 What you must do when you reach a crossing with a pedestrian on it.
9 Your car can do this on icy roads, so take care.

(Answers: Page 191)

(iii) Pedestrians and their safety

Questions

1. In what kinds of situations should I keep a particularly careful watch for pedestrians?
2. Which age groups are most at risk as pedestrians?
3. How can I spot a blind or deaf pedestrian, in order to give them extra time to cross the road?
4. At what places are there particular traffic risks to children?

Answers *(Pedestrians and their safety)*

1 In crowded shopping streets; at a bus or tram stop; near a parked milk float or mobile shop. In streets where cars are parked or stopped, pedestrians may emerge suddenly from between vehicles. Remember that pedestrians may have to cross roads where there are no crossings. Be considerate to them and drive slowly.

2 The young and the elderly. Figures show that two out of three pedestrians killed or seriously injured are either under 15 or over 60 years old. This may be because these age groups do

not judge speed well and step into the road when drivers are not expecting them to.

3 A blind person may be carrying a white stick or using a guide dog. A deaf and blind person may carry a white stick with two red reflectorised bands. Remember that deaf people may not hear your vehicle approaching.

4 The most obvious place is a school. Drive slowly near schools and look out for children getting on and off school buses. There may also be a school crossing patrol with a sign saying 'Stop — Children'. In especially busy and hazardous spots there may be a flashing amber sign which warns of a school crossing patrol ahead. Another place where children may be vulnerable is near a parked ice-cream van. Remember that they are more interested in ice-cream than traffic.

5 What should I watch for when approaching a Zebra crossing?

6 Once I have stopped, should I signal to the pedestrians to cross?

7 What are the rules about overtaking near a Zebra crossing?

8 If I am in a queue of slow-moving traffic, can I stop on a pedestrian crossing?

9 Should I give way to pedestrians who are already crossing at pedestrian crossings controlled by lights, even if the signal allows me to drive on?

Answers

5 Keep a look out for pedestrians waiting to cross over, especially the elderly, the infirm, or people with prams and children. Prepare to stop and let them cross. Remember that once someone has stepped onto a crossing you must stop and give way to them. Do not stop suddenly but signal to the driver behind that you are about to slow down and stop. Allow more time for stopping on wet or icy roads.

6 No, this is dangerous. Another vehicle may be approaching from the other direction. Let the pedestrians make up their own minds.

approach to the crossing will probably be marked with zigzag lines. You must not overtake the moving motor vehicle which is nearest to the crossing, or the leading vehicle which has stopped to give way to a pedestrian on the crossing.

8 No. You should leave all pedestrian crossings clear, so that they can be walked on.

9 Yes. Once pedestrians have begun to cross, they should be allowed to complete the crossing.

A Motto for all Road Users

Can you break the code and work out the motto? We have filled in some spaces to start you off.

Code:

```
ywmrk e vseh rixasvo mw er ibivgmwi mr gs-stivexmsr.

····· · ROAD ···W··· ·· ·· ·X··C··· ·· ·· ·········.


epp vseh ywivw qywx asvo xskixliv xs qeoi mx weji jsv epp.
··· ROAD ····· ···· ···· ········ ·· ···· ·· ···· ··· ···.
```

(Answers: Page 191)

10 What is the meaning of the signals at a Pelican crossing?

11 At a straight Pelican crossing with a refuge in the middle, must I wait for pedestrians to cross from the further side?

12 When I am turning at a road junction, should I give way to pedestrians who are crossing the road into which I am turning?

13 Must I give way to pedestrians when I am driving into or out of a property bordering on the road?

14 How should I drive past pedestrians, who are walking either singly or in large groups, on roads where there is no footpath or pavement?

15 What precautions should I take when driving past animals?

Answers

10 When the amber light is flashing you must give way to any pedestrians who are on the crossing. Otherwise the signals have the same meaning as traffic lights. A red light means stop: then comes the flashing amber light. This is followed by a green light, meaning proceed, and a steady amber light meaning stop.

11 Yes. It must be treated as a single crossing. Do not harass the pedestrians by revving your engine.

12 Yes.

13 Yes. You will probably be crossing over a pavement or footpath. Pavements are for people, not for motor vehicles.

14 Allow them plenty of room. Be particularly careful on left-hand bends and keep your speed down.

15 Drive slowly and be prepared to stop if necessary. Give them plenty of room. Do not sound your horn or rev your engine. This will only frighten them and make them less easy to control. Remember, large animals such as cattle can do plenty of harm to your car as well as vice versa. Look out for animals being led on your side of the road, especially on left-hand bends.

Also, look out for horse-riders' signals and be aware that they may not move to the centre of the road before turning right. Riders of horses and ponies are often children, so take extra care.

(iv) Breakdowns and accidents

Questions

1 If I have a breakdown, what should I do first of all?

2 What steps should I take to warn other drivers of an obstruction?

3 What should I do if something falls from my vehicle?

4 What sort of road situation will warn me that there may have been an accident?

Answers *(Breakdowns and accidents)*

1 The first priority is to get out of the way of other traffic. A car stopped in the road is an obstruction and can be dangerous, especially in poor visibility. If possible, get your vehicle off the road and keep yourself and your passengers off the road too.

2 Switch on your hazard warning lights. Place a red warning triangle (if you carry one) on the road at least 50 metres behind your vehicle, and on the same side of the road. If you carry traffic cones, arrange them on the road to guide traffic past the obstruction. The first one should be about 15 metres behind it and next

to the kerb. The last one should be level with the off side of the obstruction. At night or in fog do not stand at the rear of your vehicle as you may mask the lights.

3 Stop as soon as you can in safety, walk back and collect it or remove it from the carriageway.

4 You may see a queue of vehicles in the distance which are moving slowly or have stopped completely. You may also see police warning signs or the flashing lights of emergency vehicles.

Traffic cones moving towards the middle of the road (ref. Answer 2).

5 What should I do if I see what is possibly an accident ahead?

6 What procedure should I follow if I am the first on the scene of an accident?

7 What extra precautions should I take if the accident involves a vehicle carrying dangerous goods?

Answers

5 Slow down and be prepared to stop. Do not be distracted when passing the accident — you could cause another one.

6 (a) Turn on your hazard warning lights or set out a red triangle or traffic cones to warn other road users. Ask other drivers who stop to switch off their engines and put out any cigarettes.

(b) Arrange for the police and ambulance services to be called immediately. Make sure full details are given to them of the nature and location of the accident (particularly the number of casualties). On a motorway you will have to drive on to the nearest emergency phone.

(c) If a casualty is in immediate danger, remove him or her from the vehicle. But do not move anyone unnecessarily (see page 131). Give first aid.

(d) Make sure that uninjured people get out of the vehicles and well away from them. On a motorway get them away from the carriageway, central reservation and, if possible, the hard shoulder.

(e) Stay at the scene until the emergency services arrive.

7 You should be able to identify a vehicle with dangerous goods by its hazard information panel (see page 170). If such a vehicle is involved, you should make sure that the police and other emergency services are informed immediately about the labels and other markings. You should also keep everyone well away from the vehicle. Dangerous liquids may be leaking and dangerous dust or gases may be blown towards you. Switch off the engine and do not smoke.

(v) Level crossings

Questions

1 How should I approach and cross a clear level crossing?

2 What traffic lights are used at most modern level crossings?

3 What is an automatic half-barrier level crossing?

4 What procedure should I follow if I approach an automatic half-barrier crossing when the amber signal lights and the audible alarm is sounding?

5 Can I cross as soon as the train has gone by?

6 What should I do if I am already crossing and the amber lights and alarm start?

Answers *(Level crossings)*

1 Drive at moderate speed and cross with great care. Do not loiter on or near the crossing, and do not drive onto it unless you can see that the road is clear on the other side. Never drive nose to tail, and never stop on or immediately beyond any level crossing.

Level crossing quiz

QUESTION: Which driver is acting correctly at this level crossing?

(Answer: Page 191)

2 They have steady amber and twin flashing red lights. These must always be obeyed. If the red lights are flashing you must stop at the white line marked on the road. The barriers (if there are any) will be lowered. This means that a train is coming.

3 It has automatic barriers across the left side of the road only. These are operated by the train and are lowered automatically just before the train reaches the crossing.

4 The amber lights will be followed by flashing red 'Stop' lights, which give warning that the barriers are about to come down. Once these signals have started you must not move onto the railway. The train is nearly there and will be unable to stop. Wait at the white 'Stop' line. Never attempt to zig-zag around the barriers — you could be killed and cause a serious accident.

5 Not necessarily. If the barriers are raised, then you may proceed. But if the barriers stay down, the lights continue to flash and the audible alarm changes tone, you must wait. This means that another train is coming soon.

6 Keep going over. On no account stop and try to reverse.

7 If I am driving a large slow-moving vehicle, should I take any extra precautions before approaching a level crossing?

8 What should I do if the barriers stay down for longer than three minutes without a train passing?

9 What procedure do I follow if my vehicle stalls or breaks down on a level crossing?

10 What is an automatic open level crossing?

11 What other types of level crossing are there, and how should I proceed at them?

Answers

7 You should first of all telephone the signalman and get his permission to cross. You will find a special railway telephone at the crossing. Once you have crossed, telephone the signalman again to confirm that you are clear of the railway.

8 Use the railway telephone at the crossing to ask the signalman's advice.

9 First you must get everyone out of the vehicle and clear of the crossing. Then you should telephone the signalman and inform him of the emergency. If you have time before the

train arrives, try to push the vehicle clear of the crossing. Then phone the signalman again and tell him that the crossing is clear. If the alarm sounds or the amber light shows, get everyone clear of the crossing.

10 A crossing without gates, barriers or attendants. It has amber lights and an audible warning followed by flashing red 'Stop' lights. When the alarm sounds and the lights flash you must stop and wait. The lights will go out when it is safe to cross.

11 At a crossing with gates or barriers with skirts there may be an attendant to operate them, or they might be operated by remote control. In either case, do not cross when the barriers are down or the lights are showing. Some crossings have gates or barriers which must be opened and closed by hand. If the green light is showing, fully open both gates or barriers and cross. Then close the gates or raise the barriers again. If there are no lights, telephone the signalman for permission to cross. If there are no gates, lights or telephone, you must stop, look and listen to make sure there is no train coming. Then cross. You must always give way to trains — obviously.

(vi) Tramways

Questions

1 How should I drive on tram routes?

2 Are there any places where extra care is needed?

3 What is the procedure at tram stops?

Answers *(Tramways)*

1 You must not enter a road or lane reserved for trams. Diamond-shaped signs give instructions to tram drivers only. Take extra care on tram routes — trams can be up to 60 metres in length. The area taken up by moving trams is often shown by tram lanes which will be marked with white lines or by a different type of road surface.

2 Yes, in the following circumstances:

(a) Where the track crosses from one side of the road to the other and where the road narrows and the tracks come close to the kerb.

Watch out for separate traffic light signals which give instructions to tram drivers and other traffic.

Always give way to trams; do not try to race or overtake them.

(b) Cyclists or motorcyclists should take extra care when riding close to or crossing the tracks, especially if the rails are wet.

(c) When parking, you must not park your vehicle where it would get in the way of trams or where it would force other drivers to do so.

3 Where tram stops have platforms, either in the middle or at the side of the road, you must follow the route shown by the road signs and markings. At stops without platforms, you must not drive between a tram and the left-hand kerb. If a tram is approaching a stop, look out for pedestrians, especially children running to catch it.

Did you know that? • 3

The most enthusiastic of the early motorists were the French? In 1903, when the total world output of cars manufactured was 61,927, almost half — 30,204 — were made in France. By comparison, only 11,235 cars were manufactured in the United States and 9,437 in Britain.

Before driving licences were introduced in Britain in August, 1903, there was no age limit for drivers. The licences made 17 the minimum age for car drivers and 14 for motor cyclists. Earlier, in January 1903, the youngest reputed motor cyclist was Ernest Bond of Bristol, aged six!

All the cars in Karl Benz' first motor catalogue of 1888 were three-wheelers?

(vii) First aid in emergencies

Questions

1 What first aid should I give to an accident victim who has stopped breathing?

2 Should I try to move someone who is unconscious but still breathing?

3 How should I attempt to stop heavy bleeding?

4 Should a casualty be given anything to drink?

Answers *(First Aid)*

1 Look into the mouth and remove any obvious obstructions. Keep the head tilted as far backwards as possible. If breathing does not start again, pinch the casualty's nostrils together and blow into the mouth until the chest rises. Withdraw, and repeat regularly once every four seconds until the casualty can breathe unaided.

2 Only move them if they are in immediate danger. Otherwise leave them where they are: movement may further damage an injured limb or back.

3 Apply firm hand pressure over the wound, preferably using some clean material. Make sure there is no foreign body caught between the material and the wound. If possible, secure some padding with a bandage or length of cloth. If the limb is not broken, raise it to reduce the bleeding.

4 No, nothing at all.

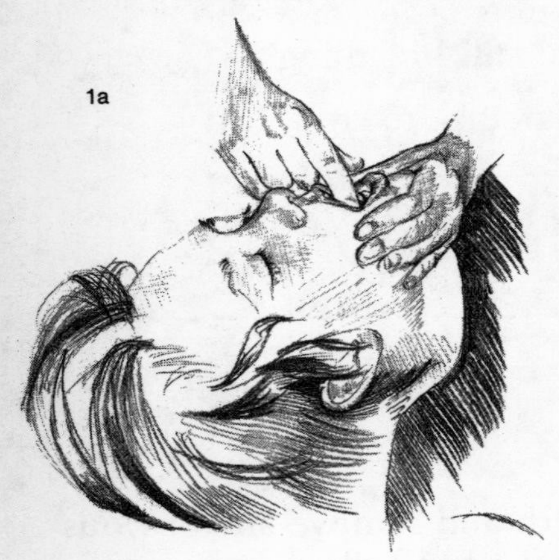

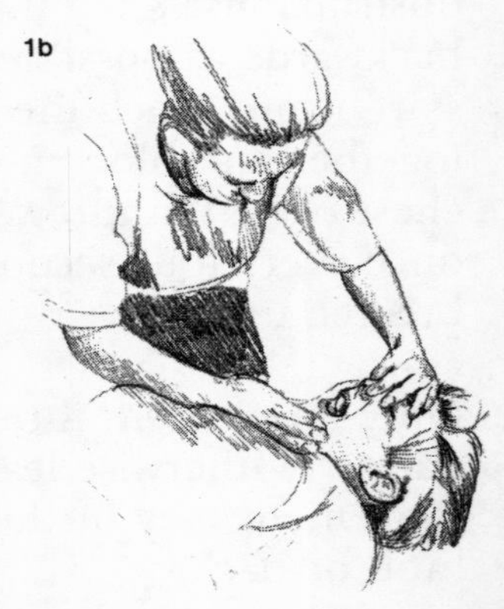

1c
2
3

(viii) Vehicle security

Questions

1 What likelihood is there of my car or its contents being stolen if I park regularly in a city street?

2 What set of rules should I follow when I leave my vehicle?

3 What further measures should I take to make my vehicle secure?

Answers *(Vehicle security)*

1 There is a one-in-four chance each year. Over 1.5 million cars are broken into or stolen every year — that's once every 20 seconds.

2 (a) Remove the ignition key and engage the steering lock.

(b) Lock the car doors.

(c) Close the windows completely (but never leave children or pets in an unventilated car).

(d) Never leave valuables or vehicle documents in the car. Take them with you or lock them in the boot.

(e) At night, park the vehicle in a well-lit place.

3 Fit an anti-theft device such as an alarm or immobiliser. Have your registration number etched on all the car windows.

Crossword Quiz

THE CLUES IN RUN ACROSS ONLY. SOLVE THEM, THEN TRANSFER THE LETTERS IN THE SQUARE TO THE GRID TO GET A MOTTO FOR SECURING YOUR CAR.

1 Where you keep your car keys while driving.
2 You should not leave your car windows like this ----, after you have parked; always park in a well ---place at night.
3 On average 180 cars are broken into or stolen in Britain every hour. How many seconds does it take to break into or steal a one car?
4 An anti-theft device (two words).
5 Colloquial term for someone who steals a car for fun (two words).
6 If you leave a case of tapes visible in your car when parking, what kind of player can a thief guess you have?
7 Where security should come in your list of priorities after parking your car.
8 You have a better chance of this when you lock valuables in your car boot.

(Answers: Page 192)

GIVE
WAY
Marton 3
HR
R
Tourist
information

LANE 7

Road Signs

It is impossible to drive far without encountering a traffic sign, a road marking or some other sort of road signal. There are, after all, more than 200 of them altogether! An accurate working knowledge of road signs is essential to succeed in your driving test. You are bound to be asked to identify some. Here is a simple pictorial quiz. Do you know them all?

Picture Quiz: Road Signs

QUESTION: What do these road signs have in common?

(Answer: Page 192)

Colour is represented in diagrams and Highway Code symbols as shown in the boxes below.

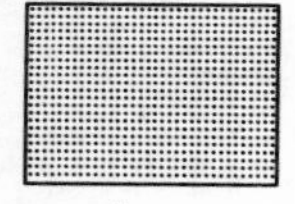

Red

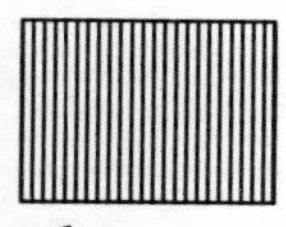

Blue

(i) Traffic sign quiz

Questions

1 What do these hand signals by police officers or traffic wardens mean?

3

4

2 What do these arm signals mean?

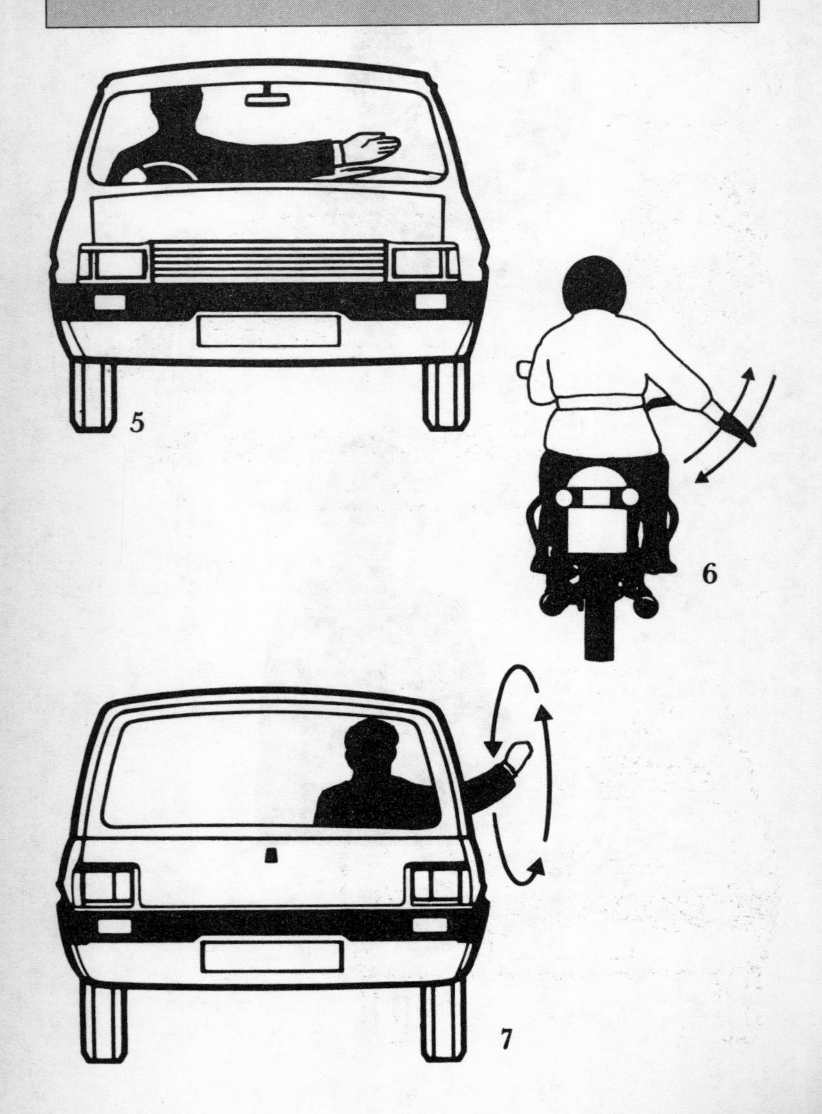

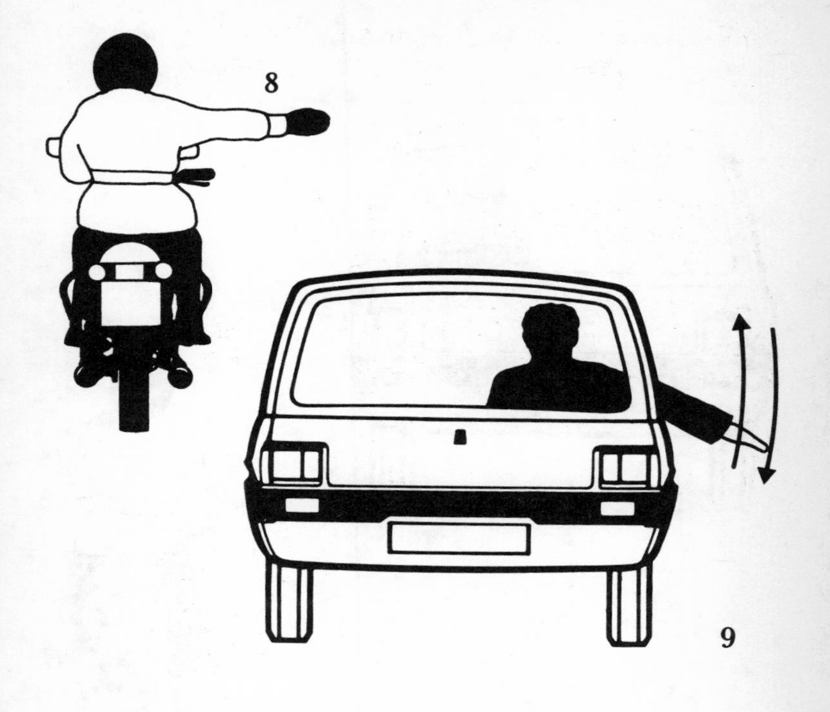

10

3 What do these direction indicator signals mean?

11

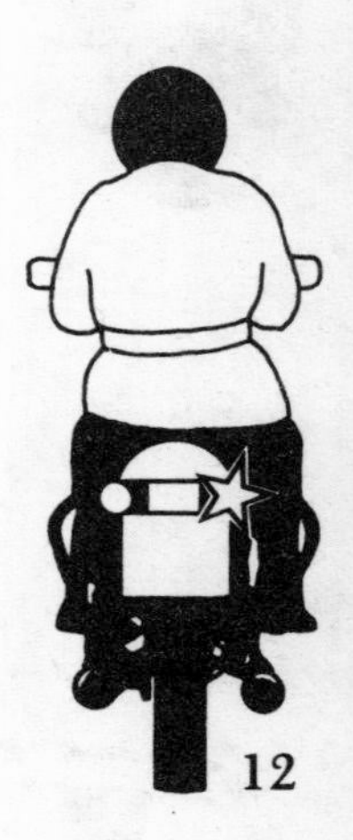

12

13

4 What do these road traffic signs mean?

14

15

16

17

18

19

20

21

22

23

24
yellow background

25
yellow background

26
green background

27

28

29

30

31

32

33

34

35

36

37

38

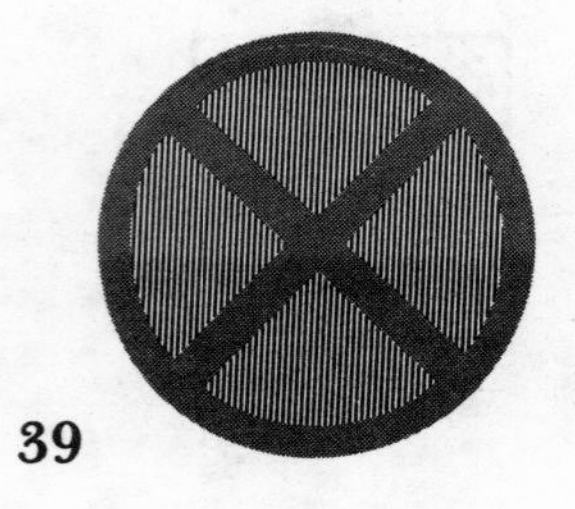

39

40

URBAN
CLEARWAY
Monday to Friday
am
8-9·30
pm
4·30-6·30

41

32
feet

42

14′-6″

43

44

45

46

47

48
yellow background

49

50

51

52

53

54

55

56

57

58

59

60

61

62

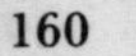

63

64

65

66

67

68

69

5 What do these direction signs mean?

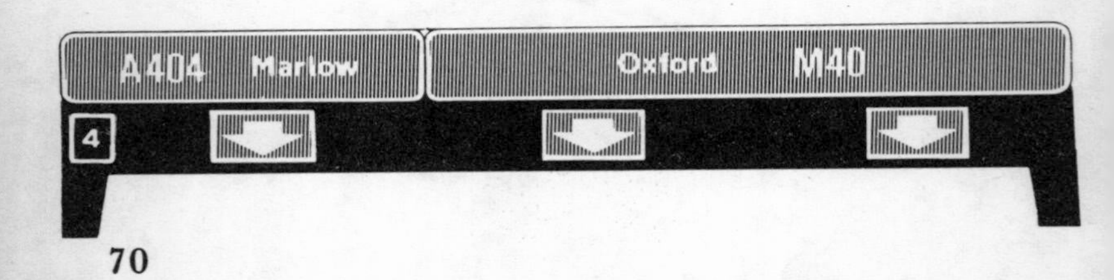

70

71

72

73

75

77

78

Controlled
ZONE
Mon-Fri
8·30 am-6·30 pm
Saturday
8·30 am-1·30 pm

74

Zone
ENDS

76

6 What do these road markings mean?

Across the carriageway:

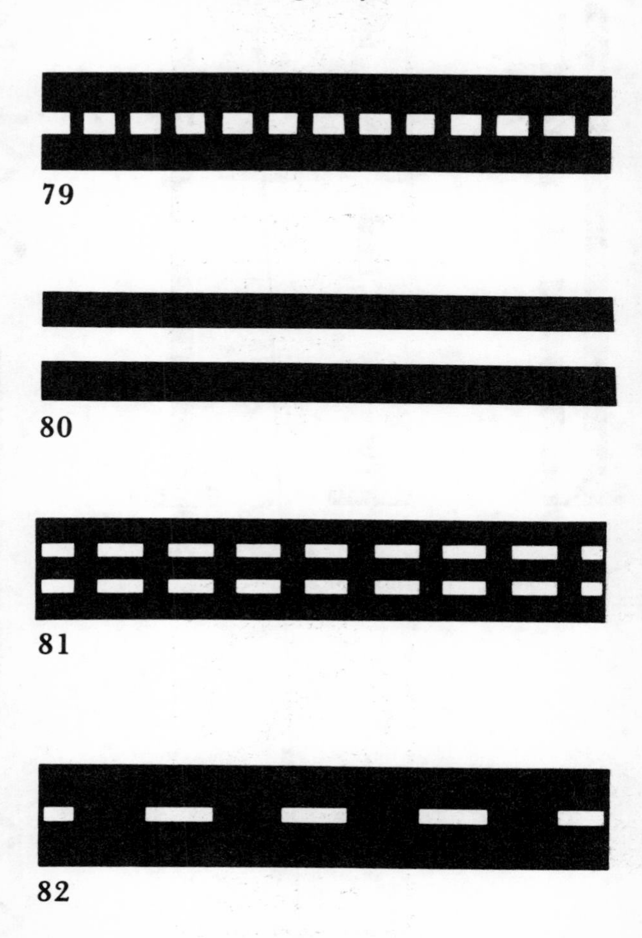

79

80

81

82

Along the carriageway:

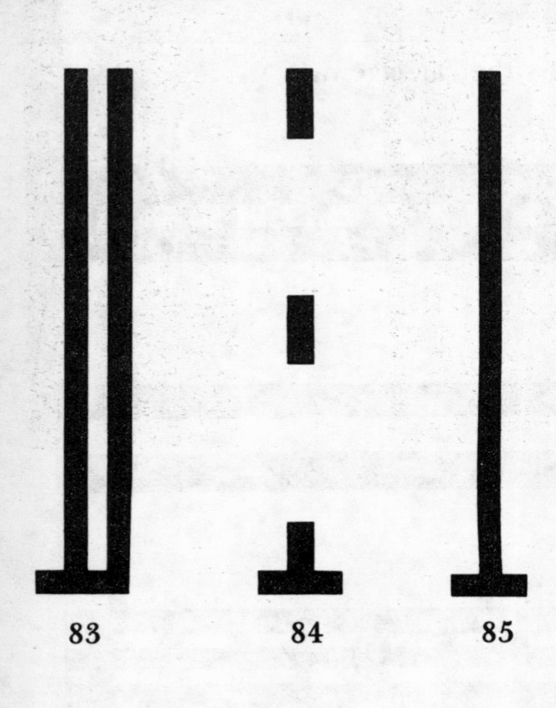

yellow lines

SCHOOL KEEP CLEAR

yellow lines **86**

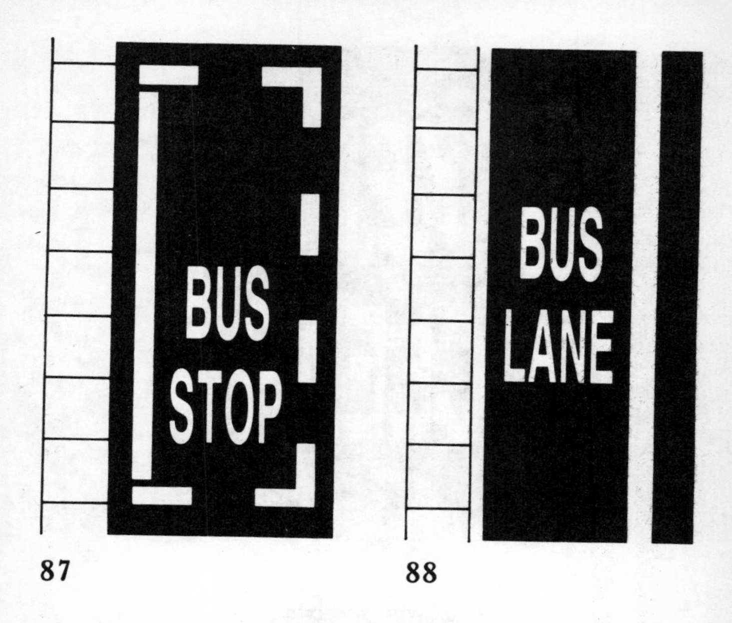

87 **88**

yellow lines along kerb

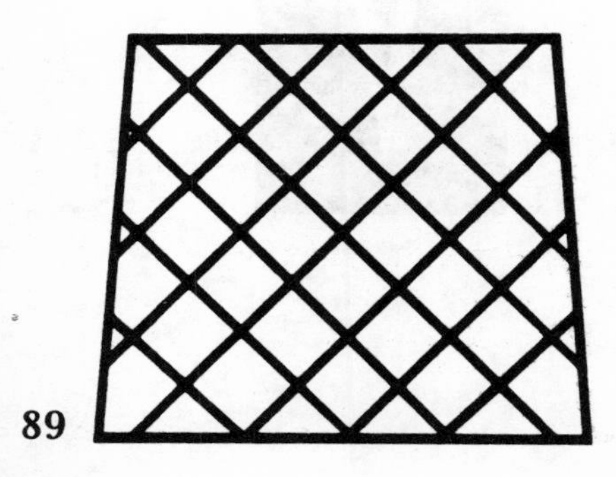

89 yellow line

90

91

7 What do these vehicle markings mean?

92

93

yellow background

94

orange background

95

96

Answers *(Traffic sign quiz)*

1 Stop: Vehicle approaching from the front.

2 Stop: Vehicles approaching from front and behind.

3 Beckoning on a vehicle from the front.

4 Beckoning on a vehicle from behind.

5 I want to turn left.

6 I intend to slow down or stop.

7 I intend to move in to the left or turn left.

8 I intend to move out to the right or turn right.

9 I intend to slow down or stop.

10 I want to go straight on.

11 I intend to move in to the left or turn left or stop on the left.

12 I intend to move out to the right or turn right.

13 I am slowing down or stopping.

14 One-way traffic.

15 No pedestrians.

16 No motor vehicles except solo motorcycles, scooters or mopeds.

17 Crossing point for elderly people.

18 Change to opposite carriageway.

19 Low-flying aircraft or sudden aircraft noise.

20 Quayside or river bank.

21 Start of motorway.

22 Parking place for towed caravans.

23 Appropriate traffic lanes at junction ahead.

24 Temporary lane closure.

25 Holiday route.

26 Ring road.

27 Cycle route ahead.

28 Overhead electric cable.

29 Two-way traffic crosses one-way road.

30 Slippery road.

31 Sharp deviation of route to left.

32 Staggered junction.

33 Roundabout.

34 Cross roads.

35 Distance to 'Stop' line ahead.

36 Keep left.

37 With-flow pedal cycle lane.

38 Vehicles may pass either side.

39 No stopping (clearway).

40 Parking restricted to permit-holders only.

41 No stopping during times shown.

42 No vehicles over length shown.

43 No vehicles over height shown.

44 No vehicles with over 12 seats (except buses).

45 No motor vehicles.

46 Give priority to vehicles from opposite direction.

47 No overtaking.

48 School crossing patrol.

49 No U turns.

50 Give way to traffic on major road.

51 National speed limit applies.

52 Mini-roundabout.

53 Minimum speed.

54 Turn left ahead.

55 Contra-flow bus lane.

56 Double bend first to left.

57 Dual carriageway ends.

58 Traffic merges from left with equal priority.

59 Road narrows on right.

60 Pedestrian crossing.

61 Failure of lights.

62 Steep hill upwards.

63 Road works.

64 Loose chippings.

65 Level crossing with barrier or gate.

66 Level crossing without barrier.

67 Height limit.

68 Wild animals.

69 Distance over which road humps extend.

70 Downward-pointing arrows mean 'get in lane'.

71 End of motorway.

72 Advisory route for lorries.

73 Recommended route for pedal cyclists.

74 Entrance to controlled parking zone.

75 No through road.

76 End of controlled parking zone.

77 Permanent reduction in available lanes.

78 Tourist information point.

79 Give way to traffic from right at mini-roundabout.

80 Stop at 'Stop' line.

81 Give way to traffic on major road.

82 Give way to traffic from right in roundabout.

83 No waiting for at least eight hours 7 a.m. - 7 p.m. plus indicated times.

84 No waiting during any other periods indicated.

85 No waiting for at least eight hours
7 a.m. - 7 p.m. on four or more days per week.

86 School entrance, keep clear, even if picking up children.

87 Bus stop: keep clear.

88 Bus lane: do not use during times indicated on plate.

89 Box junction.

90 Do not block entrance to side road.

91 Warning of 'Give way' ahead.

92 Side projection marker for overhanging load.

93 Long Vehicle marker for vehicle over 13 metres long.

94 Panel indicating that vehicle carries flammable liquid.

95 Toxic substance.

96 Spontaneously combustible substance.

•

(ii) Additional signs

Questions

1 When should I use arm signals?

2 What shape are most warning signs?

3 What colour are most signs which give positive instructions, ie those which indicate that you may do something?

Answers *(Additional signs)*

1 When your direction indicators are not used or to reinforce them. Pedal cyclists and horse riders should also use them.

2 Triangular.

3 Blue.

Picture Quiz

QUESTION: How many rules are being broken here?

(Answer: Page 192)

LANE 8

The Driver & the Law

There are thirty-one different Acts of Parliament and Regulations which are concerned with driving. To contravene any of these laws is an offence. In some cases it may even be a crime. Ignorance of the laws is no excuse. They deal with things which you either *must* or *must not* do. Many of them have already been covered in earlier chapters.

Crossword

CLUES: ACROSS

1 You commit this when you break a motoring or other law.
5 Most cars have four of this.
6 Comes after snow and requires careful driving.
8 Top colour in traffic lights.
9 Motorway driving can do this to you, so pull into a service station and have a coffee and a break.
11 Describes the lights of the car in front when driving at night.
12 Dusk, when your lights should already be on, precedes this.

CLUES: DOWN

1 Often called the "fifth" gear on more powerful cars.
2 A long way to drive.
3 Four of these hold your wheels on.
4 What the driving trest instructor is doing for you.
7 You can't start your car without it, so make sure you don't close your front door until you've checked you have it with you.
10 Shortened form of 12 across.
11 A three wheeled one is a ---cycle.

(Answers: Page 192)

(i) Pedestrians, pedal cyclists and motorcyclists

Questions

1 Am I allowed to walk on motorways or their slip roads?

2 Is it an offence to loiter on a pedestrian crossing?

3 Is a horse rider allowed to ride or lead their horse on a footpath by the side of the road if it is set apart for pedestrians?

4 Must a pedal cyclist obey the amber and red 'Stop' signals at traffic lights even when he is wheeling the bicycle?

5 If a pedal cyclist is wheeling his bicycle at night without lights, which part of the road must he walk on?

Answers *(Pedestrians, pedal cyclists and motorcyclists)*

1 No.

2 Yes. It is also an offence to obstruct free passage along a highway or to obstruct the lawful movement of a person on foot in a public place.

3 No. Nor should a horse be ridden or led onto any footway, footpath or cycle track unless there is a right to do so.

4 Yes. He is also bound to give way to pedestrians at Zebra crossings and at Pelican crossings when an amber light is flashing.

5 He must keep as close as possible to the nearside edge of the road.

6 Am I allowed to carry a passenger on my bicycle?

7 Are there any restrictions on where I can park or leave a bicycle?

8 What item of clothing must a motorcyclist wear?

9 How many passengers may be carried on a two-wheeled machine, and how must they sit?

10 Does a pillion passenger have to wear a safety helmet too?

Answers

6 No. Not unless your bicycle is constructed or adapted to carry more than one person.

7 Yes. You may not leave your bicycle on any road in such a way as to cause danger to other road users. Nor must you leave it in a place where waiting is prohibited.

8 A safety helmet of an approved design.

9 A motorcyclist must not carry more than one passenger on a two-wheeled machine. The passenger must sit astride the cycle on a proper seat securely fitted behind the driver's seat and with proper rests for the feet.

10 Yes.

Did you know that? • 4

The first road accident casualty was Mrs. Bridget Driscoll, who was run over by a Roger-Benz vehicle at Crystal Palace in London on 17th August 1896. Mrs. Driscoll was so panic-stricken at the sight of the car as it came towards her at a speed of 4 mph that she stood in its path, unable to move.

(ii) Drivers of motor vehicles

Questions

1 What documents are required for myself and my vehicle before I can drive on the road?

2 What are the laws relating to the condition of tyres?

3 What other parts of the vehicle must be maintained in safe working order?

4 Must I wear my seatbelt when I am driving?

5 What are the speed limits I must observe?

6 Is a child under 14 allowed to ride in the front of a vehicle?

Answers *(Drivers of motor vehicles).*

1 Your vehicle must be licensed (and the tax disc displayed) and properly insured. You must have a current driving licence valid for the type of vehicle you wish to drive and you must have signed it in ink. You must also have a current test certificate for your vehicle if it is over the prescribed age limit.

2 They must be suitable for the vehicle and properly inflated. They must have a tread depth of at least 1.6mm (cars, light vans and light trailers; 1mm for other vehicles) and be free from cuts and other defects.

3 The brakes and steering, the windscreen and other windows (these must be free from obstruction to vision, and must be kept clean), the windscreen washers and wipers, the seat belts and fittings, the mirrors, the horn, the speedometer, the lamps and reflectors and the exhaust system.

4 Yes, unless you are exempt from doing so or are driving a vehicle to which the law does not apply. You must also sit in such a position that you can exercise proper control over your vehicle and have a full view of the road and traffic ahead.

5 The limits are: 70 mph on motorways and dual carriageways; 60 mph on all other roads unless a lower limit is indicated by signs or by street lighting. Your vehicle may also have its own special speed limit.

6 Yes, but only if they are suitably restrained, ie wearing a seatbelt or strap which fits them properly.

7 What is the legal limit of alcohol in the blood above which one must not drive?

8 Is it against the law to use four-way flashing hazard warning lights on a moving car?

9 What must I do at night if I stop or park my vehicle?

10 What happens if I am stopped by a police officer and asked to show my driving licence, insurance certificate and test certificate, and find that I am not carrying them?

11 In what places is it illegal to park or stop a vehicle?

12 If I am involved in an accident which causes damage or injury, what must I do?

Answers

7 The limit is 80 mg of alcohol per 100 ml of blood.

8 Yes.

9 You must switch off your headlamps and leave your front and rear position lamps and rear registration plate lamps on unless you are in an area where unlit parking is allowed.

10 If necessary, you may show them within seven days at a police station of your choice.

11 Within the zig-zag lines of a Zebra crossing area; beyond the double line of studs on the approach side of a Pelican crossing; within the limits of any type of pedestrian crossing; anywhere on the road which might cause an obstruction; in a bus or cycle lane during their hours of operation; on a length of road marked with double white lines; on the right-hand side of the road at night (except in a one-way street); in an area with parking restrictions contrary to those restrictions; on a clearway; on common land more than 10 metres from a highway.

12 You must stop and give your name and address and registration number to anyone who has reasonable grounds for requiring them. Failing that, you must report the accident to the police within 24 hours.

Motoring Firsts • 4

The first car registration number in Britain was A1, issued at Christmas, 1903. The owner was Earl Russell who was so anxious to obtain the number that he stayed up all night to make sure he would be first in the queue at the registration office. Even so, Russell beat his nearest rival by only five seconds. In 1959, the A1 number plate was sold for £2,5000.

Did you know that? • 5

The first cars to travel the roads in Britain were required by law to be preceded by a man on foot carrying a red flag as a warning?

ANSWERS TO QUIZZES:

Picture Quiz: Right or Wrong? *(Page 10)*

ANSWER 1: The four pedestrians (A) and pedestrian (B) are walking correctly, away from the kerb. Pedestrian B, however, is in a better position for crossing the road since he or she can see the approaching traffic. The two pedestrians (C) are walking incorrectly; they are walking abreast and should have crossed the road, like pedestrian (D) before reaching the right-hand bend.

ANSWER 2: A "sharp bend in the road" sign.

Quiz: Who Carries What? *(Page 14)*

ANSWERS: **1** – E: **2** – A; **3** – D; **4** – A; **5** – A, B, C.

(Page 26)

ANSWER 1: A loud, two-tone horn.

ANSWER 2: The ambulance might be rushing to a road accident on the road you are using. Be prepared to stop or drive slowly if this is the case.

Five in One Crossword *(Page 32)*

ANSWERS: Across 1. Tow, 3. Ban, 5. Wet, 6. Tyre, 8.Sump, 9. Pet, 10. Tram, 11. Map, 13. Test, 14. Six.

Down 2. One, 4. Age, 6. Toot, 7. Examiner, 8. Sixty-six, 9. Post, 12. Pumps.

What Kerb Markings tell you about parking *(Page 48)*

ANSWERS:

(a) No loading outside normal working hours, that is 9.30 a.m. to 4.30 p.m.

(b) No loading at any time.

(c) Yellow lines along the edge of the road indicate the existence of waiting restrictions. Consult the nearby plate to discover what they are.

Answers contd. . . .

(Page 54)

Fortunately there are several innocent reasons why you may be asked to stop by the police. They may be checking that you are carrying the documents required by law, such as driver's licence or insurance papers. Unknown to you, a light on your car may have failed, and the police will tell you about it. Or, with so much car theft around, they may want to check that the car you are driving is really yours.

(Page 56)

ANSWER: The driver of the left-hand car.

Picture Quiz *(Page 64)*

ANSWER: Because cars and other traffic, like the lorry shown, coming the other way cannot see the parked car until they are almost on top of it. This is dangerous because it may cause oncoming traffic to swerve or stop suddenly.

Cross Quiz *(Page 68)*

ANSWERS: 1. Side road, 2. Windscreen, 3. Aerial, 4. Fog lamps, 5. Wipers, 6. Starter motor, 7. Polish.

THE LETTERS IN THE SHADED BOXES SPELL: SELFISH.

(Page 70)

ANSWER 1: This is dangerous as well as bad driving. Always overtake on the right hand side of the vehicle in front. The driver can see you coming.

ANSWER 2: On motorways, speeds are much greater than on ordinary roads. The car behind could reach the overtaking car much too quickly. Better to wait and let it pass, then try to overtake again.

ANSWER 3: Never overtake while being overtaken. It could cause a nasty accident right across the road.

Answers contd. . . .

(Page 72)

The car driver may not know that a cyclist is in front of the van he or she is about to overtake. He or she may have to drive on much further than anticipated in order to overtake both van and cyclist. If you cannot see what, if anything, is in front of the car you want to overtake, better not to overtake at all or wait until you can.

What is a Safe Gap? *(Page 78)*

ANSWER: Not really. You need more room than this to turn safely and join the line of traffic.

A safe gap is one which allows you to turn and join the traffic before the car behind you gets too close.

Roundabout Word Puzzle *(Page 90)*

ANSWERS: 1-5. Speed, 5-16. Deceleration, 17-29. No through road, 29-32. Dips, 33-36. Slip, 36-48. Parking meters.

Motorway Quiz 1 *(Page 92)*

ANSWER: (2) and (5). Pedestrians are not allowed on motorways and schools are not sited near them.

Motorway Quiz 2 *(Page 96)*

ANSWER 1: 1 — (C), 2 — (E), 3 — (F), 4 — (B), 5 — (D), 6 —(A).

ANSWER 2: Dual carriageways share characteristics with "ordinary" roads, where motorways do not. For instance, you will find traffic lights on dual carriageways, and pedestrians, too. Dual carriageways run through built up areas and contain roundabouts, whereas motorways do not.

Answers contd. . . .

Motorway Quiz 3 *(Page 101)*

ANSWER 1: There is no such thing as the fast lane. There are only driving and overtaking lanes.

ANSWER 2: On the left-hand side.

Castle Crossword *(Page 113)*

ANSWERS: 1. Train, 2. Fogs, 3. Height, 4. Snowy, 5. Rear, 6. Stays, 7. de Sacs, 8. Stop, 9. Slide.

THE LETTERS IN THE SHADED HORIZONTAL SQUARES SPELL: HIGHWAY CODE.

A Motto for all Road Users *(Page 117)*

ANSWERS: In the code, each letter is five spaces after its proper position in the alphabet, e.g. U = Y, see below.

CODE

Alphabet:	a b c d e f g h i j k l m n o p q r s t u v w x y z
Code:	e f g h i j k l m n o p q r s t u v w x y z a b c d

MOTTO: USING A ROAD NETWORK IS AN EXERCISE IN CO-OPERATION.

MOTTO: ALL ROAD USERS MUST WORK TOGETHER TO MAKE IT SAFE FOR ALL.

Level Crossing Quiz *(Page 125)*

ANSWER: Driver **A** has correctly stopped at the white line. However, he or she should never have allowed the child passenger to get out and stand by the railway line to see the train go by. Driver **B** is driving very dangerously. He is about to drive round the barrier. Driver **C** has stopped too close to the barrier after crossing the railway line.

Answers contd. . . .

Crossword Quiz *(Page 135)*

ANSWERS: 1. Ignition, 2. Open; Lit, 3. Twenty, 4. Car alarm, 5. Joyrider, 6. Cassette, 7. First, 8. Safety.

THE LETTERS IN THE GRID SHOULD READ: SAFETY FIRST LAST AND ALWAYS.

Road Signs *(Page 138)*

ANSWER: All of them warn of hazards or difficulties on the road ahead.

Picture Quiz *(page 178)*

ANSWERS: The car which has stopped on the traffic box should have waited for the box to be clear before entering it. It is unwise to run out between parked cars when trying to cross the road.

Crossword *(Page 180)*

ANSWERS: Across: 1. Offence, 5. Gear, 6. Thaw, 8. Red, 9. Tire, 11. Tail, 12. Evening.

Down: 1. Overdrive, 2. Far, 3. Nut, 4. Examining, 7. Key, 10. Eve, 11. Tri.